Active Listening

Active Listening

Michael Rost and JJ Wilson

Routledge
Taylor & Francis Group

LONDON AND NEW YORK

First published 2013 by Pearson Education Limited

Published 2013 by Routledge
2 Park Square, Milton Park, Abingdon, Oxon OX14 4RN
711 Third Avenue, New York, NY 10017, USA

Routledge is an imprint of the Taylor & Francis Group, an informa business

Notices
Knowledge and best practice in this field are constantly changing. As new research and experience broaden our understanding, changes in research methods, professional practices, or medical treatment may become necessary.

Practitioners and researchers must always rely on their own experience and knowledge in evaluating and using any information, methods, compounds, or experiments described herein. In using such information or methods they should be mindful of their own safety and the safety of others, including parties for whom they have a professional responsibility.

To the fullest extent of the law, neither the Publisher nor the authors, contributors, or editors, assume any liability for any injury and/or damage to persons or property as a matter of products liability, negligence or otherwise, or from any use or operation of any methods, products, instructions, or ideas contained in the material herein.

ISBN 13: 978-1-4082-9685-1 (pbk)

British Library Cataloguing-in-Publication Data
A catalogue record for the print edition is available from the British Library

Library of Congress Cataloging-in-Publication Data
Rost, Michael, 1952-
  Active listening / Michael Rost and JJ Wilson.
    pages cm.
  Includes bibliographical references.
  ISBN 978-1-4082-9685-1 -- ISBN 978-0-273-78611-5 (PDF)
  1. English language--Study and teaching--Foreign speakers.   2. Listening.   1. Title.
  PE1128.A2R67 2013
  418.0071--dc23
                                                              2012046155

Print edition typeset in 10/13pt Scene Std by 35

Contents

Preface

About the series

Research and Resources in Language Teaching is a ground-breaking series whose aim is to integrate the latest research in language teaching and learning with innovative classroom practice. The books are written by a partnership of writers, who combine research and materials writing skills and experience. Books in the series offer accessible accounts of current research on a particular topic, linked to a wide range of practical and immediately useable classroom activities. Using the series, language educators will be able both to connect research findings directly to their everyday practice through imaginative and practical communicative tasks and to realise the research potential of such tasks in the classroom. We believe the series represents a new departure in language education publishing, bringing together the twin perspectives of research and materials writing, illustrating how research and practice can be combined to provide practical and useable activities for classroom teachers and at the same time encouraging researchers to draw on a body of activities that can guide further research.

About the books

All the books in the series follow the same organisational principle:

Part I: From Research to Implications

Part I provides an account of current research on the topic in question and outlines its implications for classroom practice.

Part II: From Implications to Application

Part II focuses on transforming research outcomes into classroom practice by means of practical, immediately useable activities. Short introductions signpost the path from research into practice.

Part III: From Application to Implementation

Part III contains methodological suggestions for how the activities in Part II could be used in the classroom, for example, different ways in which they could be integrated into the syllabus and applied to different teaching contexts.

Part IV: From Implementation to Research

Part IV returns to research with suggestions for professional development projects and action research, often directly based on the materials in the book. Each book as a whole thus completes the cycle: research into practice and practice back into research.

About this book

Listening is now rightfully considered to be the foundation of language acquisition and communication ability. Given the importance of teaching listening in any language course, this volume brings together the most relevant research and most vital insights on listening processes and develops an innovative and engaging sequence of listening activities. The authors identify one key concept from the research, Active Listening, to be a guiding principle for educators in their design of listening activities and guidance in the most effective ways of employing them. Based on the most vital strands of listening and learning research, Active Listening is analysed in terms of five 'frames': affective, top down, bottom up, interactive and autonomous. The presentation of each frame is organised around insights into how listening ability is acquired and can best be taught. The book offers a variety of innovative and motivating classroom activities within each of these frames, together with guidance on adapting the activities to other contexts and integrating them into an overall curriculum. Readers are offered guidelines for action research projects and ideas for sharing observations and recommendations with other practitioners.

Finally, as an added feature, sample audio recordings are also provided for selected activities, available online at the series website **www.routledge.com/9781408296851**

We hope that you will find the series exciting and above all valuable to your practice and research in language education!

Chris Candlin (Series General Adviser) and
Jill Hadfield (Series Editor)

Author acknowledgments

We would like to express our gratitude to the series editors, Chris Candlin and Jill Hadfield, for inviting us to contribute to this exciting series and for challenging and guiding us during the development of this title. We also wish to thank our colleagues at Pearson for shepherding the book through production: Christina Wipf Perry, Sarah Turpie, Natasha Whelan, Sarah Owens and Sue Gard.

We would also like to acknowledge a number of professional colleagues and friends with whom we have interacted over the years. Our relationships with you have helped us formulate the framework of this book and have contributed to many of the ideas and activities in it. In particular, we wish to mention: Francis Bailey, Gerrie Benzing, Gillian Brown, Steve Brown, Antonia Clare, Jeanette Clement, Nick Dawson, Richard Day, Frances Eales, John Field, Lee Glickstein, Dale Griffee, Doreen Hamilton, Jeremy Harmer, Marc Helgesen, Kim Kanel, Tom Kenney, Salman Khan, Josh Kurzweil, Jennifer Lebedev, Cindy Lennox, Tony Lynch, Tim Murphey Yoko Narahashi, Steve Oakes, Tere Pica, Jill Robbins, Steve Ross, Joseph Shaules, Monica Silva, Rob Waring, James Wu, Mary Underwood, Penny Ur, Larry Vandergrift, Rob Waring, John Wiltshier, Junko Yamanaka, and George Yule.

Michael Rost and JJ Wilson

Publisher's acknowledgements

The publishers are grateful to the following for permission to reproduce copyright material:

Franzis von Stechow for photo p. 37; Veer/Corbis Images for photos on pp. 75 and 86; Getty Images for photos on pp. 90 and 177, Bridgeman Art Library for *The Card Player* by Cezanne on p. 170; Voicethread.com for screenshot on p. 200; Lingual.net for screenshot on p. 204; Gerrie Benzing for 'Moving Day' on p. 41.

Part I
From Research to Implications

Theoretical framework

Teachers, instructional designers and language researchers have become increasingly interested in listening. The cumulative research on listening and the growing array of listening materials available to language learners have provided us with an expanding wealth of resources, delivered in innovative technologies. At the same time, this current abundance has created a need for a fresh form of guidance. More than ever, practitioners are in need of clear principles to guide interpretation of research and to inform selection and use of appropriate resources.

We have found that one robust concept, **active listening**, can guide practitioners in identifying key principles in listening research and applying these principles in a methodical way. By active listening we are extending the connotation of 'being animated when you listen'. We are referring to a broader range of cognitive and emotional activity that could be described as 'engaged processing'.

The purpose of this introductory section is to bring together relevant research that has contributed to our understanding of this core concept of active listening and to draw key implications for practice. We will organise the range of listening research and implications into five interactive frameworks, each with an overriding focus:

Affective Frame – focus on enhancing the listener's personal motivation and involvement.

Top Down Frame – focus on deepening the understanding of ideas and making stronger interpretations.

Bottom Up Frame – focus on perceiving sounds, recognising words and syntactic structures more accurately.

Interactive Frame – focus on building cooperation, collaboration and interdependence during the listening process.

Autonomous Frame – focus on developing effective listening and learning strategies.

Each framework provides a unique perspective on the listening process as well as insights into how listening is learned and can be taught. By understanding the complementary character of these different perspectives, we can appreciate that listening development requires integration of multiple frameworks.

The five frames

Affective Frame

Some students sit silently in the classroom and feel overwhelmed and even oppressed by the listening activities the teacher presents. Because of the anxiety they feel, they tune out, have little or no engagement, perform poorly – and then *feel even worse*! Other students feel activated whenever there is a listening activity, welcoming the engagement and the challenge. They have no trouble tuning in, doing their best, and they usually make progress in listening with seemingly minimal effort.

These two types of students represent the poles of affective involvement in listening. What makes someone *want* to listen? What makes someone else *avoid* listening? Research in the Affective Frame addresses this issue and other issues related to motivation and personal engagement. Research in this framework situates the listener as the focal point of communication, an individual with affective needs and reactions, with a motivation for listening.

Key research findings

1. **The impact of motivation**
   Motivation is one of the key factors influencing the rate and success of second language (L2) learning and the level of engagement a learner is willing to undertake. Strong motivation can even compensate for weaknesses in language aptitude and for a scarcity of learning opportunities. We all know stories of amazing learners, like Mawi Asgedom, the Ethiopian refugee turned motivational speaker, who overcame daunting life circumstances and found a way to acquire a second language at the highest level, against all odds. Motivation is a cognitive force that allows the learner to maintain attention and focus. (Asgedom refers to his own motivation as a kind of 'mental karate' that provides him with a means to cut through distractions.) As a kind of fuel in the learning engine, motivation has been shown to amplify intensity of effort, intellectual curiosity and self-confidence (Aragao, 2011). Increases in motivation have also been shown to defuse anxiety and aversion to risk-taking, two factors that tend to impede language acquisition (Gardner *et al.*, 1997).

2. **The importance of the instructor**
   We have all had teachers in various subjects who have helped us 'come out of our shells' through the force of their personality, their passion for their subject, or the way they invited us to approach learning. The influence can be short-term, assisting the learner to perform better in a

specific task, or long-term, leading the learner to make strategic changes to their learning style (Williams and Burden, 1999). Classroom studies have shown that learning outcomes can indeed be influenced by several 'pedagogic agents', factors that are under the control of the instructor (Ko, 2010). One set of factors is course-specific – decisions about the syllabus, teaching materials, teaching methods, learning tasks (Ahmed, 2009). Another set of factors is teacher-specific performance – ways of showing enthusiasm for learning, ways of giving feedback, ways of building relationships with students, ways of structuring learning activities that enhance group cohesion and group support (Imai, 2010).

3. **The value of goal orientation**
Tasks pitched at the right level – not too difficult and not too easy – often lead to active engagement. Appropriately challenging tasks are likely to activate optimal levels of both emotion and cognition (Swain, 2010). Success with appropriate challenges also fuels 'internal competition' and expectations of further success, and helps learners to understand goal orientations and actively participate in goal-setting (Guilloteaux, 2007; Guilloteaux and Dörnyei, 2008). When learners understand and participate in content selection and learning goals, they will exert additional attention, effort and persistence towards achieving the goals (Williams, Burden and Lanvers, 2002.) This cyclic relationship of motivation and effort has come to be known as the **active learner hypothesis** (Oxford, 2010). Goal-oriented learners in any field, not only language learning, tend to experience an absorption that psychologists call 'flow', a deeply focused immersion in learning that contributes to a higher level of performance (Csikszentmihalyi, 2002).

4. **The effect of learner awareness**
Only truly motivated learners will be willing to face the long-term challenges involved in becoming a competent L2 speaker and listener – challenges that require a strong sense of resilience. Researchers have found that an entry point into exploring and developing a success-oriented attitude is the notion of self-awareness and bicultural identity. As learners aspire towards a positive bicultural identity, their motivation becomes a powerful force for sustaining effort in and enthusiasm for language learning (Dörnyei and Hadfield, 2013; Lamb, 2004). Many language educators advocate exploring issues of identity and social persona as the students become active users of the L2 (Norton, 2010; Morgan and Clarke, 2011). As a student develops this positive identity, he or she is much more amenable to considering new strategies – conscious ways of improving one's ability in the L2 (O'Malley and Chamot, 1990; Oxford,

2010). Meta-cognition – ways of thinking about how to learn more productively and experimenting with deliberately using new strategies – can then become a vital part of the listening instruction (Vandergrift and Goh, 2012; Rost, 2011).

5. **The power of learning styles**
   Author Barbara Prashnig has argued that people of all ages can learn virtually anything if allowed to do it through their own unique styles, their own personal strengths (Prashnig, 1998; 2006). Many L2 researchers and language educators have embraced this notion, especially given the diversity of students who undertake L2 learning. Howard Gardner's seminal work in the early 1990s established that individuals possess different kinds of intelligence and, therefore, learn, remember, perform, and *understand* in different ways (Gardner, 1991). Several distinct learning styles have been identified: linguistic (verbal), logical-mathematic, auditory (musical), kinaesthetic (tactile), visual (spatial), interpersonal (social) and intrapersonal. (Two others – naturalistic and existential/preferential – are sometimes included.) Gardner's model is a much more accessible reformulation of earlier models of cognitive learning styles, such as the Myers-Briggs Personality Inventory (Myers, 1980) and the Kolb Learning Style Inventory (Kolb, 2006; 1985). In Gardner's framework, each style involves a fundamentally different type of interaction with input (Jones *et al.*, 2009). For listening instruction, it has been proposed that engaging in multiple processing styles and consciously departing from an emphasis solely on the 'traditional' learning styles (verbal and mathematical) may have a stimulating effect on students, particularly those who have never experienced success with language learning (Lightbown and Spada, 1999). For example, numerous educators contend that kinaesthetic learning can be transformative for many students, assisting them in engaging their emotions in the learning process; kinaesthetic learning is seen as including learning through humour and laughter (which evokes positive chemical changes in the brain), drama and creative movement (Martin, 2007; Taylor, 2001; Bell, 2009).

Implications

1. Stimulating the learner's motivation is essential in promoting active listening. Active listening is triggered by affect: how the listener feels about the listening encounter, his or her level of confidence or anxiety about making an effort to listen.

2. The instructor's expertise and personal and professional qualities are vital for creating enthusiasm for learning. Selecting engaging tasks will

promote active listening: designing motivating learning tasks, generating enthusiasm towards learning, building positive relationships with students, rewarding active listening attitudes and behaviours, nurturing group support among students.

3. Offering learners choices in goal-setting and presenting appropriate challenges are likely to increase the level of motivation (additional attention, effort and persistence) required to become a proficient listener.

4. Developing learner awareness concerning the nature of language learning, and including explicit listening strategy training, is likely to improve learners' participation and increase their overall motivation for learning.

5. Providing learners with opportunities to experiment with and integrate different processing styles is likely to lead to greater motivation, more affective involvement, and better learning results.

Top Down Frame

Many students get bogged down when they are listening for an extended period, and quickly become confused and lose the train of ideas in a conversation, extended monologue or lecture. Even though they are motivated and trying hard, they have trouble grasping the main ideas and recall only muddled fragments of what they heard. Other students, even with minimal proficiency, seem to be able to latch onto the main ideas, get the point quickly, and find some personal relevance in what they have heard.

These two types of learners represent different ends of the spectrum in terms of top down processing ability. What is it that allows some listeners to tune into and recall the overall structure of what they heard, while others tend to get lost and recall only scattered words and phrases? Research in the Top Down Frame addresses this type of disparity among learners. This framework places ideas at the centre of communication, and views the listener as a problem-senser and problem-solver, obtaining cues about ideas that a speaker presents. Using interpretation of these cues, the listener then actively constructs meaning.

Key research findings

1. **The role of attention in comprehension**
   Comprehension is a complex process that involves an interaction of attention, short-term memory formation and long-term memory retrieval, all of which will be subject to individual differences. It is well-established

that when two people listen to an identical source (such as a news story) even in their first language (L1), they may vary widely in what they attend to (Perfetti and Lesgold, 1977), how efficiently they employ short-term memory (Wen and Skehan, 2011), what is relevant to their comprehension goals (Lovett *et al.*, 1999) and how accurately and completely they recall information (Miyake and Shah, 1999). One key processing difference, for example, is that some people naturally tend to pay attention to (and therefore encode in memory) the overall conceptual structure of a text as they listen, while others focus on main themes and major ideas, and still others focus on individual facts or interesting details (Hicks *et al.*, 2005). Because attention is the trigger for learning, and attention span is a predictor of learning successes, aiding students in developing attention span is seen as a vital aspect of instruction (Robinson, 2012; Juffs and Harrington, 2011).

2. **The employment of background knowledge in comprehension**
   Background knowledge is perhaps the most significant concept relating to listening comprehension. All listeners vary in their background know-ledge of a topic and invoke differing images for all of the lexical concepts they encounter in a text (Gernsbacher and Kaschak, 2003; Rost, 1990; 2006). Individual differences in accessing background knowledge (what is already known prior to listening) and encoding (what happens cognitively at the time of comprehension) are important in predicting the relative success of learners in acquiring a second language (Bransford, 2003; Johnson *et al.*, 2012). In order for understanding to take place, a listener must find common ground with the speaker of the text. Technically, this refers to sharing similar **activation spaces** in memory, a kind of mutual cognitive model (Haynes *et al.*, 2005; Levinson, 1996). Without this shared background knowledge, there cannot be adequate understanding (Binder *et al.*, 2009; Bowe and Martin, 2007; Poldrack *et al.*, 2009). When com-municating in their L1, speakers and listeners depend on activation of similar cognitive models of specific concepts, which are called **prototypes** (Rosch *et al.*, 2004). For example, shared prototypes of what we under-stand by 'a silly idea', 'a troubled childhood', or 'a creepy boss' may allow a speaker and listener of English as an L1 to communicate efficiently and empathically. However, when L2 listeners cannot rely on common prototypes or shared background knowledge with L1 speakers, they will struggle. L2 users need to construct new **neural pathways** to be effec-tive listeners: to understand new concepts in the L2, to find interest in what L1 speakers are talking about, and to empathise with their emo-tional states (Haynes and Rees, 2005; Churchland, 2006; Alptekin, 2002; Benet-Martínez and Lee, 2009).

3. **The effect of pre-listening tasks**

   Comprehension is possible only when a degree of expectation is present before listening. It is now accepted that comprehension is fundamentally a matter of understanding context and therefore expecting meaning, rather than waiting for meaning to emerge (Ross, 1975; Tannen, 1993.) It is also established that cognitive expectations ('schemata') that are activated prior to listening significantly influence what is understood and how well it will be remembered (Bartlett, 1932; Chafe, 1977; Rumelhart, 1980; Lustig et al., 2001). Several particular types of activating expectations are known to be assistive in comprehension and retention: lexical priming (previewing concepts and key words in advance) (Chang, 2007), visual priming (previewing images related to the input) (Ginther, 2002), pre-listening questions (prompts about what will be asked while you listen or after you listen) (Graesser and Person, 1994), and advance organisers (previewing the overall rhetorical organisation) (Herron et al., 1995; Chang and Read, 2007).

4. **The benefit of post-listening tasks**

   While pre-listening activity influences comprehension, activities the listener undertakes during and after listening will significantly influence retention. It is known that, in many conditions, active tasks while listening have an additive effect on processing of input; productive tasks provide a more potent encoding effect. Additional positive effects, such as improved recall, are observed when active while-listening tasks such as note-taking, are paired with post-listening tasks (Armbruster, 2000). In particular, the most effective post-listening tasks seem to be those that encourage interactive review (such as reviewing a script or a set of notes) (Carrell et al., 2002; Kiewra et al., 1995) and **appropriation** of the input (such as doing further research and giving a presentation on a related topic) (Donato, 2004; Thornbury, 2002).

5. **The multi-modal nature of language processing**

   We often think of listening as involving only the aural channel, but nearly all listening involves attention to signals in multiple channels. In live face-to-face communication as in multimedia text processing, the listener must attend to multiple layers of input, including the verbal system (written, auditory and articulatory verbal codes), the non-verbal system (visual and symbolic content), and the physical context (including people and objects referred to) (Cross, 2011; Gruba, 2006). In actual listening then, the listener processes large amounts of information simultaneously and makes complex configurations of mental representations while listening (Paivio, 2007). From this integration of aural and visual input (some of which may be confusing) and cognitive input (what the listener already knows), the listener attempts to construct meaning (Hymes, 1964; 2009). Some researchers propose new models for understanding multimedia processing, involving a 'dynamic image schema' that provides the listener with additional means of organising and analysing information (Qiu and Huang, 2012). Revisiting the act of listening from this multi-modal perspective entails new systems of understanding meaning that extend beyond language only, and systems of instruction that integrate multiple modes of processing (Cope and Kalantzis, 2012; Anstey and Bull, 2006).

Implications

1. Giving learners opportunities to focus on ideas, rather than simply on language, and focusing on building comprehension of complex ideas is central to listening development.

2. Developing a curiosity about new ideas and expressions in the L2 is an important aspect of active listening. Listening activities that promote **mediation** (with the learner's ideas in the L1) and appropriation add to the listener's strategies for understanding challenging input.

3. Pre-listening is an essential platform for listening development. Activities involving activation of images, concepts and organisation structures before listening are likely to be effective at promoting top down processing.

4. Post-listening is a vital aspect of listening development, particularly for reviewing listening strategies, for developing memory and for enhancing appropriation of content and skills.

5. Listening involves paying attention to multiple channels, not only the oral channel, so it is useful to highlight all sources of knowledge with which we construct meaning. Providing learners with controls over multiple input channels (such as subtitles on/off) helps them isolate sources of information as they listen.

Bottom Up Frame

Some students in the class may appear to be natural communicators – they speak fluently and they seem to tune in to others easily. It seems that they are listening very well – *until* someone checks what they actually heard. They may be mishearing key words and completely missing or ignoring complex grammatical phrases. Although they are managing to get by through compensating with their top down listening ability, they are suffering at some level from a lack of bottom up processing skills.

Is it possible to teach this type of student how to pick up on more of what is actually spoken, to hear the language more completely and more accurately? Understanding of research in what we call the Bottom Up Frame can certainly help in formulating an approach. This framework focuses on the language that is used during the process of communication, the precise audio signals that are available to the listener and how they are perceived and decoded.

Key research findings

1. **Attention lapses during perception**
   A major cause of difficulty in second language listening is due to attention failures when an unfamiliar sound sequence is perceived. The attempt to process unfamiliar or unexpected sounds produces what is known in neurolinguistics as a P-400 effect, a temporary processing lapse in the auditory cortex (Dien *et al.*, 2010). (This processing 'blip' typically occurs about 400 milliseconds after an unfamiliar sound is perceived, hence the name.) Encountering and recovering from frequent lapses of this nature puts greater stress on the listener, and often leads to communication breakdowns (Rost, 2006), to debilitating effects on motivation to continue listening (Graham, 2006), and to an undue reliance on inferencing processes (Field, 1998). To counter these perceptual lapses, it is essential that the L2 listener becomes familiar with the **phonemes**, **intonation patterns**, and the **phonotactic system** of the L2. For children learning a second language, this process of familiarisation tends to progress without problems as the child learns to differentiate L1 and L2 phonemes accurately from the 'puddles of sound' in the environment (Monaghan and Christiansen, 2010; Singleton and Ryan, 2004; Kraus, 1999). However, for the adolescent or adult L2 learner, the process of acquiring the L2 phonology is typically quite problematic. Until the L2 learner begins perceiving the sounds of L2 accurately, the learner's L1 'intrudes' in the perception process. As a result, the learner 'attenuates' incoming sounds, effectively hearing the closest sound in his or her L1 instead of the actual uttered sounds of the L2 (Cutler, 2011; Cross, 2009). Because this perceptual constraint is an evolutionary feature of the human auditory system (we are supposed to identify with one 'tribe' by the time of adolescence), the learner must actively retrain the auditory cortex to prevent 'intrusion' of the L1 (Cutler, 2011; Hasan, 2000).

2. **Adjusting to fast speech phenomena**
   Speech rate is a common problem for L2 listeners. L1 listeners' comprehension is typically unaffected by speeds of up to 250 words per minute,
   whereas even advanced L2 listeners typically reach optimal comprehension at 130 words per minute (Buck, 2001). Difficulties occur not so much
   because of speed itself, but because the L2 listener is not prepared
   for **fast speech phenomena**. Even at normal speaking rates (about 180
   words per minute), 'fast speech' (or connected speech) phenomena occur
   continuously (and relentlessly for the L2 listener). Consonant sounds are
   systematically **assimilated** – co-articulated or fused (and sometimes **elided**
   entirely) – rather than individually articulated. For instance, when you say
   the phrase 'fast speech', you are likely to assimilate the consonant cluster
   in the middle and just say something like 'faspeech'. Vowel sounds are
   systematically reduced – formed in central rather than peripheral areas of
   the mouth. For instance, when you say the word 'independent' in normal
   speech, you are likely to 'reduce' the first 'e' sound and the final 'e' sound
   (in unstressed syllables) to a central 'schwa' sound. When you grow up
   hearing a language, you learn to decode these differences automatically.
   However, for L2 learners, this 'automated' process is anything but automatic! It is now known that L2 learners tend to use the rhythmic system
   from their L1 to decode L2 speech. This strategy, except with very closely
   related languages, is generally counterproductive, and leads to faulty
   word recognition and grammatical parsing (Broersma and Cutler, 2008;
   Cutler, 2012; Indefrey and Cutler, 2004; Altenberg, 2005).

3. **The need for automatic word recognition**
   Word recognition is the essential operation in bottom up processing. L2
   learners typically may know a word in isolation, but cannot recognise it in
   connected speech. For example, a learner may clearly know lexical items
   like *vegetable* or *comfortable* or *first of all*, but in connected speech, may
   hear 'vech tipple', 'come to Paul' or 'festival'. A major aspect of bottom up
   listening training is coming to recognise words automatically (without
   conscious attention or backtracking). In order to facilitate **automatic word
   recognition**, L2 learners need multiple exposures to the same vocabulary
   items (an advantage of 'narrow viewing' in the same genre), multi-channel
   exposures (audio and text), and **spaced repetitions** (encountering the
   word at regular intervals). While various types of practice aid learners
   in acquisition, research shows that a key to increasing recognition of
   vocabulary is in intensifying the amount of engagement while listening
   (Schmitt, 2010; Nation, 2008; Segalowitz, 2008; DeKeyser, 2009; Bialystok
   amd Craik, 2010; Sydorenko, 2010; Rodgers and Webb, 2011).

4. **The supportive role of parsing in speech recognition**

   Parsing speech (assigning words and phrases in terms of grammatical relationships) is a key part of bottom up processing. Many grammar signals may be elided in speech, so the listener has to anticipate and fill in much of the grammar to keep up with the speaker. Because of the importance of grammar, some teacher trainers advocate a **grammar noticing approach** to supplement listening instruction (VanPatten *et al.*, 2009). A few different tacks have been researched. Aural training that employs enriched input (highlighting target structures) or processing instruction (using form-focused tasks) has been shown to be effective for increasing awareness of grammar while listening (Field, 2008; Rost, 2006; De Jong, 2005; Wilson, 2003).

5. **The integration of linguistic and paralinguistic signals**

   Decoding speech involves more than recognising sounds and words. **Paralinguistic signals** (the sounds that ride on top of the words) contribute to the overall signal that is produced by the speaker (Ohala, 1996; Brazil, 1995). This level of speech can modify and amplify potential meaning. For example, it can provide cues about emotion (the speaker's attitude towards topic and specific words), salience (relative importance of new information), connections (relationships between words and phrases) and staging (intonational and melodic patterns to indicate desired effects on the listener) (Cheng *et al.*, 2005; Eckert and McConnell-Ginet, 2003).

Implications

1. Learning to hear the new sounds and sound patterns in an L2 is essential for listening, but requires sustained effort and motivation, particularly for adult learners. Targeted instruction in L2 sound perception is essential.

2. To become proficient listeners, L2 learners need to become comfortable with fast speech. Active training in the sounds and sound patterns of the L2 (in regular, small doses) is necessary for all learners, particularly those who have learned the L2 first through reading.

3. Word recognition is the most important bottom up listening process. Listening instruction that focuses actively on word recognition will improve learners' overall listening ability.

4. Grammatical parsing is a key skill in listening. Listening instruction that includes grammatical processing, or noticing of grammatical forms, will be of value for most learners.

5. Actively attending to paralinguistic signals can assist learners with aural perception in the L2. Instruction that raises awareness of the role of paralinguistic signals will add to the learners' repertoire of listening strategies.

Interactive Frame

There are always some students in a class who have scored at the highest levels on reading and writing sections of proficiency exams but who struggle mightily with aural communication. Some may feel embarrassed at the gap between their reading and listening skills. They're reluctant to interact and to ask questions because their 'weakness' may be revealed. Because they don't interact willingly, they tend to get ignored, and get fewer opportunities to develop a much needed ability.

This type of issue is explored in what we call the Interactive Frame of listening research. This framework views the listener as a participant at the centre of communication. In this type of research, meaning is viewed as interactive, co-constructed between the listener and the speaker.

Key research findings

1. **Creation of learning opportunities**
   Affective involvement of the listener during an interaction is a key determinant of the quality of understanding he or she experiences (Ushioda, 2010). Affective involvement can be an antidote for anxiety and aversion to speaking. Listeners with greater involvement tend to be more motivated to participate, to be open, to feel more confident in their abilities, to reveal more of themselves, and thus have a more satisfying interaction and a more valuable learning experience (Yang, 1993; Aniero, 1990). On the other hand, listeners who perceive themselves as discounted or marginalised tend to adopt a low action orientation. This leads to greater perceived social distance with the speaker, which reduces motivation to interact (Villaume and Bodie, 2007; Ford *et al.*, 2002).

2. **Communication strategies and quality of interaction**
   The goal of active listening in an interaction is to develop a degree of interpersonal solidarity. This involves developing strategies to 'tune in' to the motivations of the speaker, to provide appropriate participation signals, and to actively guide the interaction towards a desired outcome. Various measures and scales have been developed in psychology to monitor communication strategies and listener strategies in interaction (Rost and Ross, 1991; Nakatani, 2010). These scales provide barometers

for acquisition of listening abilities in social, professional and therapeutic settings, and can be used as a means of feedback and assessment (Luoma, 2004; Ducasse and Brown, 2009).

3. **Listener response as central to interaction**
   One of the core skills in interactive listening is **backchannelling** – providing appropriate displays of comprehension, empathy and readiness for subsequent turns. It will include combinations of brief verbal and semi-verbal signals (*ahhs* and *hmms*, for example), laughs, chuckles and postural movements, such as nods. While all languages have some form of backchannelling, the style and frequency differs from culture to culture. Learning to backchannel using L2 norms can have a beneficial effect, in terms of perceptions of fluency and proficiency in L2 listeners. An important aspect of backchannelling is signalling when understanding has gone awry, and initiating some form of **repair** (White and Burgoon, 2006; Wolf, 2008; McCarthy, 2010; Moore, 2011; Chalhoub-Deville, 1995; Lazaraton, 2002.)

4. **Interaction and pushed output**
   Interaction plays a vital role in providing opportunities for negotiation of comprehensible input, an important channel for language acquisition (Gass and Mackey, 2006). (This position is known in second language acquisition research as the **interaction hypothesis** (Long, 1996)). Oral interaction tasks that present a need and opportunity for negotiation of meaning are an essential aspect of language development. Some interactive tasks also provide opportunities for **pushed output**. 'Modified output', speech that the learner modifies following feedback from a listener, 'pushes' learners to produce more accurate, appropriate, complex and comprehensible forms (Weinart, 1995; Krashen *et al.*, 1984; Swain, 1993). Swain has claimed that modified output benefits L2 development because 'learners need to be pushed to make use of their resources; they need to have their linguistic abilities stretched to their fullest; they need to reflect on their output and consider ways of modifying it to enhance comprehensibility, appropriateness and accuracy' (1993, pp. 160–1). Pushed output tasks, when joined with listening input, can be especially effective in 'forcing' learners to reconstruct ideas using precise vocabulary and grammatical structures (de la Fuente, 2002).

5. **Learner initiation of interaction**
   Research on classroom interaction has shown that there is a reflexive relationship between pedagogy and interaction in the L2 classroom. How the teacher interacts with students and how the students interact with

each other are foundational elements of instruction, particularly in classes emphasising development of oral communication. Indeed, it can be argued that all pedagogy is translated into interaction (Seedhouse, 2004). Analysis of classroom interaction can provide insights into how learners and teachers interpret the pedagogy, and lead the teacher towards making informed decisions about how to influence classroom discourse to promote learner-initiated interaction (Richards, 2006) and how best to utilise instructor feedback to promote learning (Lyster and Mori, 2006).

Implications

1. Stimulating interactive tasks can increase affective involvement while creating opportunities for listening development.

2. The main strategy in interactive listening is consciously creating solidarity with the speaker. Feedback on learners' use of listening strategies during interactions is likely to improve their proficiency.

3. Listener response is a vital communication skill. Backchannelling is the fundamental mode of listener response and should be practised, in its various forms, as part of listening instruction.

4. Interactive listening tasks are essential for language acquisition because they require negotiation of meaning and pushed output. Collaborative problem-solving tasks and input reconstruction tasks can promote this kind of development.

5. Classroom learning entails interaction: all pedagogy is realised through some form of interaction. By investigating the 'interactional architecture' of classroom conversations, teachers can make informed decisions on designing tasks to promote learner-initiated interaction.

Autonomous Frame

We have all seen some students soar in listening and communication skills – make quantum leaps during our course. Though we may want to attribute at least some of their success to our teaching, it is more often the case that these students are doing something valuable *outside of class* that is propelling them forward.

As important as classroom teaching is – for fuelling motivation, for scaffolding learning tasks, for developing learning strategies, for providing professional feedback – language learners are not likely to make substantial and sustained progress without independent learning *on their own*. The Autonomous Frame of research is focused on independent learning. In particular, research in this framework is concerned with how learners come to be self-directed and productive in their independent learning activities.

Key research findings

1. **Necessity of supplemental learning opportunities**
   Autonomous listening is essential to language development, as language classrooms seldom offer sufficient listening opportunities for sustained progress. Different forms of hybrid learning – involving varying proportions of classroom and self-access instruction – have been compared. Particularly in EFL environments in which contact with live speakers of the target language is rare, autonomous online learning is an attractive option for developing listening ability. Autonomous learning outside the classroom appears to be most effective when the teacher provides monitoring and periodic assessment, as well as a means of actively integrating the two modes of learning (Little, 2008; Ushioda, 2009; Tong and Tong, 2010).

2. **Connecting to social contexts**
   Autonomous learning is not only concerned with finding more opportunities for learning, but with connecting to social contexts outside the classroom (Block, 2003; Firth and Wagner, 1997; Milne, 2007). The most basic application of this idea is identifying learning opportunities outside the classroom that connect with classroom tasks (Crabbe, 2007) and becoming aware of English-learning possibilities in the immediate environs (even in EFL contexts), what Lynch calls 'WOT' (the 'world out there') activities that serve as a springboard for classroom learning (Lynch, 2009). A deeper aspect of becoming autonomous involves becoming interdependent with speakers of the target language community (Toohey, 2000; Lynch, 2009). One aspect of community-based learning may be **service learning**, engaging students in meaningful service to their schools and communities. Through this kind of engagement, students not only learn more language in real-world settings, but also experience an authentic connection to people around them (Russell, 2007; Moser and Rogers, 2005).

3. **Integrating new technologies**
   New technologies have provided, and will continue to provide, increasing opportunities for autonomous listening outside of the classroom, for the integration of listening into multimodal learning and the development of multi-literacies (Goodwin-Jones, 2012; Cope and Kalantzis, 2012). Authentic materials incorporating listening (audio and video) have become increasingly accessible to learners. Research has established that

additive online-based listening instruction enhances students' listening ability (Romeo and Hubbard, 2010), and has a positive effect on their attitudes towards online language learning (Puakpong, 2008). Online forums, discussion groups and social networking have also enhanced opportunities for listening. One study has shown that listening behaviours account for almost three-quarters of the time learners spend in online networking activities. Cluster analysis has identified three distinct patterns of behaviour and indicates that online listening is a complex phenomenon and a substantial component of students' participation in online learning (Wise *et al.*, 2012; Choe, 2011).

4. **Developing instructor support in autonomous learning**
   Language advising is now part of the expanded role of the modern teacher. In addition to classroom teaching and coaching, instructors now are expected also to assist learners in finding learning paths outside the classroom. This role extends beyond recommending suitable learning resources outside of class, to counselling learners in creating their own learning goals (Mynard and Carson, 2012). Deciding on the appropriate amount of support (and more is *not* necessarily better) is a crucial aspect of planning for out-of-class learning: the key is in adjusting cognitive load so that learners have the optimal level of challenge (Ishikawa, 2012; Benson and Chik, 2010). For instance, there are varying views about the use of captioning and help menus, vocabulary glosses and translations when teaching with multimedia. One view is that the additional support will increase context, reduce cognitive load and improve comprehension (Clark *et al.*, 2006; Jones and Plass, 2002). A parallel view is that any support that allows the learner to keep listening is a valid tool (Rendaya and Farrell, 2011).

5. **Developing learning strategies**
   The development of learner strategy inventories and guides for the good language learner has helped teachers understand the possible paths for learners towards autonomous learning. In particular, teachers are learning how to promote **self-management strategies**: organising and planning your own learning, monitoring and managing your learning, evaluating the outcomes of your learning (Oxford, 2011). Learning strategies that are specific to listening are those that the learner consciously chooses in an effort at improving his or her planning, focusing attention, monitoring, evaluating, inferencing, elaborating, collaborating and reviewing (Vandergrift, 2011; Rost, 2011). (See Appendix 1 for a full list with examples.) It appears that learners who explore, develop and adopt learning strategies will often perform slightly better on specific listening tasks

(Lynch, 2009). Learners who consciously adopt strategies to improve also make more sustained progress towards overall proficiency (Vandergrift and Goh, 2012; Cross, 2009; White, 2008; Ushioda, 2008; Vandergrift and Tafaghodtari, 2010).

Implications

1. Autonomous listening is essential to sustained progress in listening. To maximise effectiveness, some form of teacher monitoring and assessment is necessary.

2. Out of class listening can help learners expand their identity as L2 users. Social context listening tasks that encourage planning, self-discovery and reflection provide students with the challenges and benefits of real-world learning.

3. New technologies and particularly internet technologies allow for individualisation of learning resources and tasks. Various support technologies enable learners to access a range of authentic sources and communities.

4. The instructor's role in autonomous learning is active, as is the student's role. Instructor tasks include planning, advising, mentoring, providing coaching and feedback.

5. Learner strategy instruction will enhance the learners' experience with autonomous listening. In particular, self-management strategies need to be promoted.

Summary

Part I has outlined key research strands that have informed the learning activities we present in Part II. We have organised the research into a set of five frames, each reflecting a particular aspect of listening:

Affective – focusing on **motivational processes**

Top Down – focusing on **inferential processes**

Bottom Up – focusing on **perceptual processes**

Interactive – focusing on **interpersonal processes**

Autonomous – focusing on **strategic processes**

Because of the multi-dimensional nature of listening, we propose that *all five frameworks are of equal importance*, and *all five should be drawn upon* in language instruction. You will find that there is an overlap in the research findings, learning principles and listening processes for each frame. For example, listening activities in all frames will draw upon motivational ideas in the Affective Frame, to some extent. Similarly, because top down and bottom up processes are complementary, you will notice a kind of synergy between these two types of activities in the Top Down and Bottom Up Frames. In a similar vein, because *all* instruction is realised as some form of interaction, you will observe aspects of the Interactive Frame throughout activities in the other frames. Finally, principles of the Autonomous Frame, especially the development and use of active listening strategies, apply to specific aspects of activities in all of the other frames. We have grouped particular activities in each of the frames to represent the kinds of activities that enable students to develop the focus of that frame.

Over all five frames, the development of listening strategies is important for progress. We define a listening strategy as a conscious attempt to improve one's listening comprehension or listening task performance. We have identified eight categories of active listening strategies: Planning, Directed Attention, Monitoring, Evaluating, Inferencing, Elaborating, Collaborating and Reviewing. By purposefully attempting to use particular strategies during the listening activities (we list target strategies for each activity in the Aims section), students can become more active – mentally and physically – and more engaged. We have included a separate table of active listening strategies in Appendix 1 as a reference.

Part II
From Implications to Application

As we outlined in Part I, listening can be profitably researched from various perspectives. Each of these perspectives offers unique insight into the nature of listening and to the manner in which 'active listening' is developed.

We suggested five frameworks for applying principles of listening research to teaching:

Affective – focusing on motivational processes

Top Down – focusing on inferential processes

Bottom Up – focusing on perceptual processes

Interactive – focusing on interpersonal processes

Autonomous – focusing on strategic processes

Because of the multi-dimensional nature of listening, we propose that *all five frameworks should be used* in language instruction. As all five frameworks are directed towards the ultimate goal of helping the learner become a proficient and self-sufficient L2 listener, there is no ideal proportion or sequence for using the activity frames.

There are 10 representative activities outlined for each frame, 50 in all. The activities are presented in a schematic fashion:

- **Introduction** outlines the purpose of the activity and highlights how it engenders 'active listening'.

- **Aim** states the learning goal of the activity, from the perspective of the learner: what the learner is intended to notice, experience, practise, or learn.

- **Level** indicates the general proficiency level (Beginner, Intermediate, Advanced) of the students who will benefit from the activity. Adaptations of input and procedure, of course, can be made to alter the suitability for different levels and ages of students.

- **Time** suggests an approximate amount of class time required to complete the basic activity. Suggested options or variations will increase the time required.

- **Materials** section of the activity provides a ready-made or easily accessible (free, online) short list of resources to be used for the activity. The materials suggested are listed in approximate order of difficulty, from 'easiest' (most appropriate for beginner level) to 'most challenging' (most appropriate for advanced level).

- **Preparation** outlines any rehearsal or preliminary work you, the teacher, need to do before you use the activity in class.

- **Procedure** lists the essential steps of the activities, dividing the steps into 'before listening', 'listening' and 'after listening'. Some steps, labelled 'optional' in the procedure sequence, may be omitted without altering the basic activity.

- **Variation** provides an alternate activity, often more experimental, that addresses the same learning aims as the main activity.

- **Comments** offers personal observations from the authors on why the specific activity has been particularly interesting or effective in our own teaching.

- **Worksheet** furnishes a ready-to-use support for the activity, which may be a duplicable task sheet to guide the students, an actual text or set of texts for the teacher to use in the activity, or a table of language forms that can be used by the students as part of the activity.

- **Research links** provides links to resources for the teacher who wishes to probe the learning principles inherent in the activity, or to explore what other teachers have done with similar activities.

Frame 1
Affective Frame

Introduction

The **Affective Frame** deals with the all-important area of motivation. Probably more than any other issue in teaching, motivation tends to dominate our attention: How can I motivate my students? How can I get them to relax? How can I help them feel a bit more connected to the others in the class? Are my students' expectations realistic? Does my own enthusiasm influence my students? Does my choice of activities affect their motivation? Do activities have to be enjoyable to be motivating for them? What do I do if the students aren't motivated?

All of these questions about motivation arise in the teaching of listening. We know from our experience that *the choice of activities does indeed influence learner motivation* – particularly the type of input, the amount of preparation and guidance, and the style of feedback given by the teacher. The 10 activities in this frame are all designed to address issues of learner motivation by *making the learners more active* – giving them tangible actions to perform (to take their minds *off* comprehension), interacting freely with classmates and the teacher in order to collaborate more fully, and creating a playful atmosphere that allows them to relax and enjoy the learning process. With activities in this frame, it is useful to have learners focus on developing strategies of planning, evaluating and elaborating. (See Appendix 1 for a list of active listening strategies and examples of specific strategies the students can try out.)

Ten illustrative activities

The ten activities presented in the Affective Frame are:

Activity	Description	Focuses on improving motivation	Focuses on instructor involvement	Focuses on goal orientation	Focuses on learner awareness	Focuses on developing learning styles
A New Skill	learning a skill through listening	●		●		●
Fly Swatter	listening to select the correct word and 'swat' it					●
Pinch and Ouch	using drama techniques to focus on sounds				●	
Photo Album	listening to personal stories using picture cues	●	●	●		
Emotional Scenes	tuning in to the emotions of characters	●			●	●
Guided Journey	using visualisation to develop listening and motivation	●		●	●	●
Listening Circles	giving supportive feedback to classmates		●		●	●
Wrong Words	listening for mistakes in transcriptions of song lyrics	●				●
Finish the Story	using imagination to complete a story	●	●	●		●
Punchline	understanding and evaluating jokes	●		●		●

(Darkened boxes indicate research links to the activities.)

Activity: A New Skill

Introduction

This activity involves teaching the students a skill, thereby removing the focus from studying language. Even if the hobbies presented are distant from the students' experiences, the teacher's enthusiasm usually keeps them with you for the crucial listening input stage. During the activity, the students are totally focused on meaning, and are often unaware that they are involved in listening practice. Once the students have caught some of the energy from the instructor's model, the second stage of the activity tends to be highly enjoyable as it places the student at the focal point of the class. The activity is particularly appropriate for kinaesthetic learners.

Aim	• increase motivation to listen
	• use hands-on learning
	• develop these active listening strategies: directed attention, comprehension monitoring, multimodal inferencing, world elaboration, seeking clarification
Level	Beginner +
Time	20 minutes

Materials

Whatever your hobby is, try to bring in: something visual; something that the students can touch; and an example of what the hobby looks like in practice. The latter could be a video of you hang gliding, a loaf of home-made bread, an internet download of someone doing yoga, even a performance of a song on the guitar.

Here are some filmed examples of people instructing others in new skills:

http://www.youtube.com/watch?v=JK0DTF9Edtk (how to make naan bread)

http://www.youtube.com/watch?v=XL-MrZKU3Js (how to play guitar for beginners)

http://www.youtube.com/watch?v=e0rSmxsVHPE (how to meditate)

http://www.youtube.com/watch?v=kCt1bmSASCI (how to juggle)

http://www.youtube.com/watch?v=NBkmzBdS-gE&list=PL5865505CCA3E0D7C&index=6&feature=plpp_video (boxing – how to jab)

See Howcast videos (www.howcast.com) for a good selection of 'How to' clips including tech and gadgets, food and drink, sports and fitness, etc.

Preparation

Prepare to teach your hobby (or at least some of the basics of your hobby) to the students.

Procedure

1. (before listening) Pass around the pictures or objects associated with your hobby. Then tell the students about it from a very personal point of view. Say how long you have been doing it, who taught you, what it means to you, and why you love it.

2. (listening) Move on to the practical stage of the lesson. This involves a 'How to . . .' approach. Demonstrate the activity, getting students physically to perform it if possible, even if they are only miming. Explain the stages involved in the activity, showing them which bits are difficult. If appropriate, the students can actually play the game being described, do the activity or hobby, or create something.

3. (after listening) Recap by eliciting the stages of the hobby and writing them on the board (or using other projection media).

4. Tell the students they will explain their own hobby (this could be in the next lesson). Give them time to prepare.

Variation

Get the students to complete a worksheet (see below) as a first listening comprehension stage.

Comments

Using the same principles as this activity, language classes have embarked on cooking lessons, made box kites, learned how to play rugby, built sculptures made of Lego, and so on. We have found that most students respond well to the challenge of learning a new skill or hobby. We have also found that, besides learning a new skill, we learned a great deal about the students: their hobbies, their confidence when placed in the position of 'expert' and their teaching style.

Research links

Krashen, S. and Terrell, T. (1983) *The Natural Approach: Language Acquisition in the Classroom.* Hayward, CA: Alemany Press.

Shannon Hutton's article on helping kinaesthetic learners succeed: http://www.education.com/magazine/article/kinesthetic_learner/.

The Natural Approach, a resource site for teachers: http://naturalway.awardspace.com/index.htm.

Worksheet

Complete the following:

Name of hobby/skill

Place/time to do it

Materials

Time needed to learn it

Special abilities

Activity: Fly Swatter

Introduction

This activity involves short bursts of intensive listening. The students listen to descriptions of target words and win points for their team by 'swatting' the correct words. Because the activity involves a comprehension race, the students will use the skill of inferring in order to 'swat' the answer more quickly than their opponents. The friendly game-like competition involved tends to be very motivating. This activity helps develop the interpersonal learning style, without creating too much of a demand for verbal output.

Aim
- use games in the classroom
- increase enjoyment of listening
- develop these active listening strategies: selective attention, comprehension monitoring, linguistic inferencing, predictive inferencing

Level   Beginner +

Time   10 minutes

Materials

Two plastic fly swatters.

Sources of vocabulary:

Course books: many have word lists either at the back of the book or at the end of units, e.g. *Straightforward, Total English, English File, English in Common*, etc.

Vocabulary books: Watcyn-Jones, *Test Your Vocabulary* series; Redman, McCarthy, O'Dell, Mascull, *et al. Vocabulary in Use* series; Yates, *Practice makes Perfect: English Vocabulary for Beginning ESL Learners*; Schmitt, Schmitt and Mann, *Focus on Vocabulary* series.

Preparation

1. Choose a number of target words/phrases to write on the board. These are the words that you will describe or allude to. Try to have at least twice as many words/phrases as there are students in the class.

2. Prepare to talk about the words/phrases on the board without naming them. You could give a straight description or talk about the topic, which the students then identify.

Procedure

1. (before listening) Write the target words and phrases on the board in random order and not in lines (see sample below).

2. Put the students into two teams. Each team has one chair a metre from the board. The chair is facing away from the board. There is a fly swatter on each chair.

3. The students in each team take turns to sit on their team's chair and hold the fly swatter.

4. Explain the following: you will talk about one of the items on the board. As soon as they can guess which item is being talked about, the students stand up quickly, turn around, and swat it. The winner is the person who swats the correct word/phrase first. Each student gets only one swat, so if someone makes a mistake, he or she has to sit down again. Other team members are not allowed to shout out the answers; they must remain silent.

5. (listening) Do a practice run. Two students sit in the chairs. The other team members stand on their side of the board. Talk at a steady, natural speed, adjusting your delivery for the level. Emphasise that it's a race to swat the correct word/phrase.

6. Once the practice is finished, explain that every time someone on your team wins, the team gets a point. Keep a running tally of the points.

Variation

Teams can shout out the answers. This will rapidly descend into chaos, but it's part of the enjoyment.

Comments

The activity sounds like a child's game and, indeed, children love it. However, we have used it successfully with adults of all ages and in different types of class: medical students (reviewing medical terminology), business students, and other specialised classes. The combination of movement and competition seems to appeal to most students.

Research link

Wise, D. and Forrest, S. (2003) *Great Big Book of Children's Games*. New York: McGraw-Hill.

Sample Image

Two-word verbs and collocations placed randomly on the board:

wake up   get up

have breakfast

leave home   go to work

have lunch   get home

have dinner   take a shower

go to bed

Activity: Pinch and Ouch

Introduction

Drama techniques can be very effective for teaching active listening and for promoting interpersonal learning styles. 'Pinch and Ouch' is a basic acting principle that derives from dramatic training: we can't give an 'ouch' until we feel an actual 'pinch'. This activity involves ambiguous dialogues, which require active listening, paying attention to verbal cues (what the person says), paralinguistic cues (pacing, pausing, loudness, intonation, manner of articulation) and to non-verbal cues (facial expressions, gestures, body language). The active listener will attempt to tune in to all of these cues.

Aim
- raise awareness of phonological cues
- develop spontaneity in responding
- develop these active listening strategies: emotional monitoring, noticing attention, speaker inferencing, creative elaboration

Level   Beginner +

Time   15 minutes

Materials

Use the short dialogues on the worksheet or locate similar ambiguous short dialogues and situations for improvisation.

Creative Drama offers accessible ideas for improvisations, theatre games and skits. www.creativedrama.com/theatre.htm

Preparation

Read over the skits you are using, noting places where intonation and non-verbal cues will impact meaning.

Procedure

1. (before listening) Pass out the worksheet with A and B lines, or use a projected image of the worksheet that all of the students can see. Give the students a couple of minutes to read the dialogues silently.

2. Explain that you're going to ask the students to imagine a situation for each dialogue: Where are the speakers? Who are they? What is their relationship? How do they feel in this conversation?

3. (listening) Read the first dialogue with your choice of inflection and intonation, taking both parts. (Or you may pre-record the dialogues with two actors.) Ask the students to think of situations. Encourage as many guesses as possible. There is, of course, no single correct answer.

4. (after listening) Have students work in pairs to practise the dialogues. Focus on listening: the B speaker should listen closely to how A says the line before responding.

5. Have the students choose one of the dialogues and create a 10-line extension. Volunteer pairs can act out their dialogues in front of the class.

Variation

Emotional states. If students have difficulty describing possible situations, ask them what emotion they hear in your (or the speaker's) voice: You're _____ (angry, suspicious, jealous, sad, happy). Adaptations and other ideas for younger learners can be found at Drama Resources: http://dramaresource.com/games

Comments

We have found incorporation of drama techniques in the classroom to be very valuable. Some of our inspiration has come from Augusto Boal's *Theatre of the Oppressed* (Boal, 1979), which uses drama techniques to highlight issues of social justice. We've also been inspired by Richard Via's approach, embodied in his groundbreaking book, *English in Three Acts* (Via, 1976), and which he so generously shared through his teacher workshops.

Research links

Maley, A. and Duff, A. (2005) *Drama Techniques in Language Learning*, 3rd edition. Cambridge: Cambridge University Press.

Teacher Lingo. Role play ideas: http://teacherlingo.com/blogs/makeadifference/default.aspx.

Worksheet

Ambiguous dialogues

	What's the situation? A is . . . / B is . . .
A: You've changed. **B:** I have.	
A: I'm leaving. **B:** I'll miss you.	
A: Help me. **B:** I can't.	
A: May I help you? **B:** Yes, how much?	
A: Is this your first time here? **B:** No, it's not.	
A: You've kept me waiting. **B:** Ah.	
A: You look sick. **B:** Do I?	
A: It's been a long time. **B:** I know.	
A: It's time to go. **B:** Not yet.	
A: Are you busy this weekend? **B:** Um, no.	
A: Is Christine there? **B:** No, she's not. Who's calling?	
A: How are you? **B:** I'm OK.	
A: Stop it. **B:** Make me.	
A: I'm sorry. **B:** It's all your fault.	
A: What are you doing? **B:** What does it look like?	

Activity: Photo Album

Introduction

This activity is very personal and can often promote group cohesion and enhance the spirit of collaboration in the class. Because of the personalised aspect, the activity tends to be motivating for students. The activity deals with real world and personal information, so it promotes a natural curiosity which is key to active listening. While extrinsic factors (tests, exams, bureaucratic information) can be 'compelling' as motivators, this activity uses the students' intrinsic interest in the topic and the speaker. The activity uses visual input, which appeals to visual learners.

Aim • develop motivation to listen

• develop these active listening strategies: self-management, selective attention, speaker inferencing, multimodal inferencing, personal elaboration

Level   Beginner +

Time   20 minutes

Materials

Teacher's photos; students' photos.

Preparation

1. The day before this lesson takes place, ask the students to bring in photos of friends and family, if they do not have photos available on a mobile hand-held device.

2. Plan to speak about a selection of your photos (see sample visual and script).

Procedure

1. (before listening) Pass around a selection of photos of people who are close to you. Get the students to speculate about the people in the photos.

2. (listening) Stick the photos on the board so students can see them. If necessary with big classes, use blown-up photos or large slides. Tell the students you are going to describe the people in the photos, and your

relationship with these people, without indicating which one you are talking about. The students' task is to listen and guess who you are talking about.

3. After each description, pause and ask them who you were talking about. Confirm or deny their guesses.

4. (after listening) Tell the students that now it's their turn to describe photos of friends and family. Give them several minutes to prepare. Provide new vocabulary as appropriate.

5. In pairs or groups, they do the activity.

Variation

- Provide the input twice: first, at natural speed, and second, at a reduced speed.

- Tell the students you will describe each picture in turn but one of the descriptions will contain a lie. The students' task is to work out what the lie is. This generally motivates students to listen attentively because of the pleasure involved in 'catching' the lie.

Comments

Many classes will be interested in their teacher's private life. When this is *not* the case, the motivational factor may be a funny story about one of the people in the photos, or a funny photo.

Research links

James Abela ELT site ideas: http://www.jamesabela.co.uk/advanced/realia. html.

Mumford, S. (2005) 'Using creative thinking to find new uses for realia', *The Internet TESL Journal*, 11, pp. 1–3. http://iteslj.org/Techniques/Mumford-Relia.html.

Ur, P. (1984) *Teaching Listening Comprehension*. Cambridge: Cambridge University Press.

Sample Visual and Script

Photo courtesy of Franzis von Stechow

Audio File 1: Photo Album

Please visit www.routledge.com/9781408296851 for an audio recording of this transcript.

This is my godfather. He looks somewhat fierce in this official photograph, and that's not surprising because he was rather fierce in his public life. He was a Professor of Literary Philosophy in his native Germany. I guess this was taken around 40 years ago, by which time he was already becoming established. He was one of the pioneers of something called Reader Response Theory, and I remember his students were all terrified of him. Apparently, he could be rather withering in his comments if they didn't catch on quickly enough. However, I remember him with enormous fondness. He was great with children – he spoiled me rotten – and when I visited him as a young adult he would take me for spectacular meals in beautiful restaurants and talk to me about writing and books and great authors. He was a lovely man.

Activity: Emotional Scenes

Introduction

This activity involves listening to short emotionally charged monologues and dialogues, with the focus on empathising with the speaker. This type of activity helps develop the inferencing strategy because it allows the listener to use context, paralinguistics and kinesics to infer meaning. It also involves the elaborating strategy, particularly personal elaboration, or connecting with one's personal experience to build meaning. Because a major part of listening involves tuning in to non-verbal and intuitive elements of meaning, students will realise that they can comprehend a great deal, even if they do not understand all of the words a speaker says. This type of activity appeals to the intrapersonal learning style, in that it values reflection and intuition.

Aim	• raise awareness of the role of emotion in communication
	• work with authentic audio and video clips
	• develop these active listening strategies: personal elaboration, world elaboration, linguistic inferencing, contextual inferencing, creative elaboration
Level	Intermediate +
Time	20 minutes

Materials

Short texts of emotional monologues or speeches that you will read aloud.

Preparation

Rehearse the short monologues, noting any special inflections or emphases you may wish to add (see samples).

Procedure

1. (before listening) Tell the students you are going to read some short texts that contain a lot of emotion. Elicit some common terms for emotion: angry, afraid, happy, sad, surprised, etc.

2. Write these questions on the board: Who is the speaker? How old is he/she? What is the situation? What is happening now? How does he/she feel?

3. (listening) Read the text aloud slowly, with dramatic effect. Pause from time to time to make eye contact with the students. Repeat the reading, with the same emotional nuances.

4. (after listening) When you finish reading, pause, and point to the questions on the board or repeat them orally. Have the students answer the questions in pairs.

5. Elicit answers from the students, listing as many plausible possibilities as they come up with.

6. Have the students reflect on how the speaker shows emotion: through their voice, intonation, choice of words, physical actions. Ask the students to consider how they understand the emotion of the actors in this integrated way, how they use their intuition to sense the intention of the speaker.

7. Ask the students to choose one of the scenes to rehearse and act out. Can they recreate the emotion of the speaker? Coach them to use variations in volume, pausing, chunking of phrases, and emphasis on particular words to create different emotional effects.

8. As a follow-up, have the students write their own emotional monologue. This could be an actual childhood scene or a fictional scene. Then they read their monologues, and the other students identify the feelings and emotions involved.

Variation

Movie scenes. Choose a short, emotionally charged scene from a film or TV show. First, watch with the sound off. How much can the students understand of the scene? Who are the characters? What is their relationship? What is happening? What emotions are the characters feeling? Next, play the scene with the sound track on. How much more can they understand?

Possible sources:

- *Gray's Anatomy*. Izzie talks with her patient. http://www.youtube.com/watch?v=wFyoWWzYERw

- *Comrades*. Two soldiers reminiscing. http://www.youtube.com/watch?v=h3KyEKfuvOE

- *Good Will Hunting*. Will defends his friend. http://www.youtube.com/watch?v=ymsHLkB8u3s

Comments

We have found that many aspects of active listening that we learn in our L1 can be adapted to L2 teaching. One of these aspects is **reflective listening**, something that listeners do when trying to help the speaker deal with a difficult emotional state. It is very similar to what the applied linguist Deborah Tannen (2001) calls rapport-talk. In reflective listening, the central question for the listener is, 'How does the speaker feel in this situation?'

Research links

Bar-On, R. (1997) *Bar-On Emotional Quotient Inventory: User's Manual.* Toronto: Multi-Health Systems.

Carroll, J.B. (1965) 'The prediction of success in foreign language training' in Glaser, R. (ed.), *Training, Research, and Education*, pp. 87–136. New York: Wiley.

Carroll, J.B. (1993) *Human Cognitive Abilities: A Survey of Factor-analytical Studies*. New York: Cambridge University Press.

Sample Short Monologues

Audio File 2: Emotional Scenes

Please visit www.routledge.com/9781408296851 for an audio recording of this transcript.

Moving Day (young child reflecting on family deciding to move)

We're moving away. To another city. In another town. About a million miles away from here. My Mom has a new job that makes her really happy and my Dad says he can do his job anywhere.

That's great for them, but what about me?

This is my house. This is my tree. I love this tree. I've climbed it. I've had picnics under it and, once, when I was angry, I even kicked it. Now I'll never see it again. It isn't fair.

(Monologue reproduced with kind permission of Gerrie Benzing. Gerrie teaches and directs children's and teenagers' theatre classes at the Cultural Park Theatre in Florida, USA.)

Graduation Walk (college graduate reflecting on her family's past)

I'll be thinking about you today. I'll be thinking about you when I walk up the aisle. I'll be thinking about you when they hand it to me. My diploma. This is for you. This is what you gave me. A chance. Moving to this country to give me a better chance at life. A chance. Pushing me to go further than I thought I could. A chance. Showing me that there's nothing I can't overcome. I'll be thinking about you today.

Activity: Guided Journey

Introduction

This activity presents a framework for using visualisations as part of a listening–speaking task. It also puts an emphasis on the role of the instructor in furnishing a supportive environment for the activity. Creative visualisation is an exercise for activating the imagination and allowing listeners to tap into their memories. Visualisations can be used to relax, to generate new ideas, and to develop constructive attitudes. When the content of the visualisation is positive and constructive, these exercises can be invigorating and supportive of the students' learning goals.

Aim
- practise relaxation while listening
- strengthen a positive identity as a language learner
- develop these active listening strategies: persistent attention, emotional monitoring, creative elaboration, visual elaboration, summarisation

Level Intermediate +

Time 10 minutes

Materials

A 2–3 minute visualisation script that you will read verbatim. For ideas involving identity and language learning, see Dörnyei and Hadfield (2013).

Preparation

Create your own visualisation script or use the one in the sample. If you create your own, use these steps:

1. A short (one minute or less) 'induction' creating a relaxing and intriguing natural scene.
2. A metaphorical story (two minutes or so) in which the student is the actor, including some 'positive choices'.
3. A conclusion or 'awakening' that brings the student back to the present situation.

Procedure

1. (before listening) Preview the key vocabulary. (See the items in *italic* in the sample script.) This preview will allow students to begin to form images for the key events in the story.

2. (listening) Have the students relax and close their eyes. They may wish to rest their heads on their desks.

3. Read the script slowly, in a soothing voice. Repeat or rephrase parts, as needed, but don't worry if the students don't understand every word, idea, or image.

4. (after listening) Give the students a minute to 'come back' to the normal classroom environment.

5. Have the students work in pairs to tell their story. The listener should listen for:

   - What is your gift?
   - Why is it special for you?
   - Are you grateful for this gift?
   - How can you use this gift to help you?

6. Ask your students to reflect on the activity: Did they like it? What parts did they enjoy? Do they feel it's a useful listening activity?

Variation

Imagined story. Have the students close their eyes. Narrate a story, slowly and with a lot of redundancy. Intersperse the story with several open-ended questions such as: You see (a person). What does (the person) look like? / You enter (a room). What do you notice? What do you hear? / (The person) approaches you. What happens next? At the end of the imagined story, have the students work in pairs to recount the story they experienced, using as much detail as they can.

Comments

We have used this kind of visualisation in teacher-training workshops because it helps teachers connect their own experience with the theme of the work-shop. One variation is the 'meet your mentor' theme. The listener re-imagines meeting a person who has been very influential in their life. The listener asks questions of their mentor and receives their answers. Following this exercise, the teachers share the messages they have received from their mentors, and explain how this guidance has influenced them in their teaching.

Research links

Tomlinson, B. (2001) 'The inner voice: a critical factor in L2 learning', *Journal of the Imagination in Language Learning and Teaching*, 6, pp. 1–10.

Sample Visualisation Script

Audio File 3: Creative Visualisation

Please visit www.routledge.com/9781408296851 for an audio recording of this transcript.

The Gift

(introduction)

Close your eyes. Relax. I'm going to tell you a story. Listen. Imagine. Notice what you see, what you hear, what you feel. Just notice, and enjoy the story.

(story)

You are walking down a quiet *path* in the *forest*. You hear birds singing and you feel a soft breeze on your skin. You turn on the path and you notice *an old house* up ahead. What does the house look like? You walk up to the house and see a large *wooden door*. You open the door to *a large room*. What does the room look like?

You look around the room and in the *corner of the room* you notice *a large box*. You walk over to the box and open it carefully. Inside the box, there is *a gift*, just for you. You pick up the gift and *examine it*. What does it look like? What does it feel like? This is your gift, *something special*, just for you.

You hold the gift, and turn around. You *appreciate* the gift, and you know the gift will help you in your life. You go back outside and notice the *blue sky* and feel *the warm breeze* all around you. It is a beautiful day. You're happy that you *discovered* this path, you're glad that you found this house, you're pleased that you *received* this gift. You feel *grateful* and you know that you will use this gift well. Now . . . think of a way that you can use this gift . . . to help you become a better language learner. Can you think of a way?

(awakening)

Now it's time to come out of the forest, to come back. Come back slowly. Slowly. Open your eyes. You feel *refreshed* and *relaxed* and ready to continue with . . . your *journey*.

Activity: Listening Circles

Introduction

Active listening involves attempting to connect with the speaker, to develop a 'listening relationship'. This goes beyond showing that you are paying attention to the speaker's message; it demonstrates that you are supporting the speaker's efforts to communicate. This activity presents an explicit framework for practising supportive listening, and appeals very directly to the interpersonal learning style. The key attitude to engender in this activity is honing in on the *best aspects* of the speaker's effort and giving *only positive* feedback. What makes this activity work is a *positive regard* towards the speaker.

Aim
- develop a supportive learning environment
- experience the benefits of group support
- learn some supportive listening gambits
- develop these active listening strategies: persistent attention, speaker inferencing, backchannelling, emotional monitoring, mediating

Level Intermediate +

Time 3 minutes per student

Materials

A list of personal questions that your students can easily talk about. Some examples:

- What is something you are passionate about?
- What is your favourite place to relax?
- Who has had the most influence in your life?
- What has been a pivotal event in your life?

Preparation

Prepare a list of personal topics that your students can choose from.

Procedure

1. (before listening) Arrange the group in circles, preferably no more than eight people per group. (Optional) Have the students introduce themselves to one another, and allow a minute for small talk.

2. Tell the class they are going to practise 'listening support'. Explain the procedure:

   - Everyone will have a turn to speak for one minute. You can choose a topic from the list.
   - When you speak, don't rush. Make eye contact with one listener at a time. Try to keep eye contact while you speak. Move your gaze from one person to the next. (You may want to model this as you speak.)
   - Listen carefully to each speaker. Lean forward and maintain 'warm' eye contact with the speaker. Keep a neutral expression. (You don't need to smile.)
   - You (the teacher) will keep time. Give each speaker one minute after you say, 'Begin'. With 5–10 seconds left, say, 'Please finish up.' Then at one minute, say, 'OK, thank you.'
   - Speakers, when your turn is over, please wait in front of the group. Accept your applause. Then listeners can call out, individually, their 'support feedback'.
   - At the end, the listener will give 'support statements' only. You can use the 'listener feedback' samples. Read some of the examples aloud, and have the students say them along with you.

3. (listening) Begin the activity by naming the first speaker, or use a random selection process. Use a timer (most mobile phones have a stopwatch) to keep the activity fair for all students.

4. At the end of each speaker's turn, continue with the feedback process. You can add your own feedback as well.

5. (after listening) (Optional: Do this step in the students' L1.) Discuss the effectiveness of the activity: Did you enjoy this? What was the part you liked most? How can we change this activity for the next time? Assign the topics for the next time you will try this activity, so students can have an opportunity to plan ahead.

Variation

Silent turn. If students are particularly nervous about speaking, you can do this activity with the first turn of each speaker as a silent turn – no words, only eye contact with the listeners. This may actually help students understand that a lot of their nervousness isn't about speaking itself, but rather about 'being seen' and making silent connections with the audience.

Comments

We learned the outlines of this activity from Lee Glickstein (2007), who has helped us adapt his approach for L2 learners. Although the task structure seems uncomfortable at first, this activity framework works well once your students relax and experience the benefits of having a supportive listening environment. If the teacher consistently promotes 'positive regard' for all students, it will help develop a classroom culture that will support more relaxed participation and better listening.

Research links

http://www.speakingcircles.com/Articles/Principles/EssenceApp.html.

http://www.newconversations.net/sevenchallenges.pdf.

http://www.youtube.com/watch?v=-dpk5Z7GIFs.

Sample Listener Feedback

Type 1: Speaker Essence

- You're _____!
- You're very _____!
- _____!

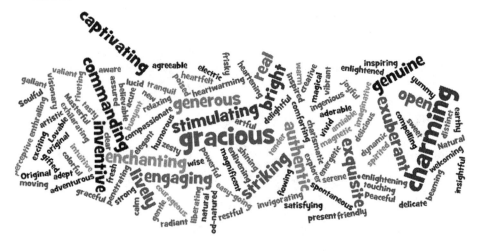

Type 2: Speaker Content

- I learned a lot!
- That was very _____ (interesting, informative, useful, original, uplifting).

Type 3: Speaker Language

- Your voice is _____ (beautiful, nice, easy to listen to).
- I love your _____ (pronunciation, intonation, vocabulary, grammar, way of talking).

Activity: Wrong Words

Introduction

Songs are an incredibly rich resource for teaching listening. Music brings an extra dimension to language classes – it may bring a great tune, rhythm, a story, a beautiful voice or heartfelt emotion, all of which can be highly motivating for students. The use of songs containing rich lyrics can also benefit language acquisition; grammar, vocabulary, phrasings and phonology can be reinforced through repetition, and memory can be aided through the rhythmic aspects of music. Furthermore, students can feel more fluent – both as listeners and as speakers – when working with songs. This activity is based around a song and it involves a game-like process of listening for 'mistakes' in the song lyrics.

Aim	• listen for detail
	• learn language points from music
	• improve concentration and memory
	• develop these active listening strategies: double-check monitoring, problem evaluation, selective attention
Level	Intermediate +
Time	20 minutes

Materials

- An audio recording of a song that you can play in the classroom. The lyrics should be appropriate for your students in terms of level, subject matter and tone (watch out for obscenities).

- A worksheet with the words of the song changed to 'soundalikes'.

Possible sources:

http://grooveshark.com/

http://www.playlist.com/

http://www.youtube.com/

To find lyrics, see:
http://www.songlyrics.com/
http://www.lyrster.com/
http://www.azlyrics.com/

Preparation

Find a song with simple lyrics. Obtain a copy of the lyrics and change some of the words to 'soundalikes'. Change content words such as nouns and adjectives rather than words such as prepositions and articles. For example, from *Someone Like You* by Adele: 'I heard that you sat in town' (instead of 'I heard that you settled down'); from *We Found Love* by Rihanna: 'Yellow time runs in the light' (instead of 'Yellow diamonds in the light'). Make at least 10 changes to the lyrics and make sure the changes do not come too closely together, i.e. only one change per line.

Procedure

1. (before listening) Do some pre-listening work with the song: write the title on the board. Ask if the students know the song. Add six or seven key words from the song. Ask if they can predict what type of song it is (e.g. a love song).

2. (listening 1) The students listen to the song and note any phrases that they understood.

3. Hand out the song lyrics with the soundalike 'mistakes' and explain that the students will need to correct 10 mistakes. Give the students a few minutes to do this.

4. (listening 2) They listen again, correct the mistakes and compare their corrections with a partner.

5. A final listening task is either: (a) simultaneous silent reading and listening, or (b) the students sing along while reading the corrected lyrics.

Variation

• Show music videos instead of only listening. Besides added interest, videos provide the benefit of allowing students to read the singer's lips, a useful aid to understanding the lyrics.

- Lip synch. In this light activity, students move their lips in synchronisation with the recorded song. Any well-known songs can be used for this, but repetitive, high-energy songs, especially those with chorus parts, work best. Examples:

'Don't Stop Believing' (*Glee* theme song) http://www.youtube.com/watch?v=xIoSTbPt_PI

'The Locomotion' http://www.youtube.com/watch?v=C5OoQadZTPk

'Hey, Soul Sister' http://www.youtube.com/watch?v=EeGDRSWB46w

Comments

Authentic songs can be difficult listening material; to understand song lyrics in the target language is, for many students, a landmark in their development as language learners, a pleasurable measure of progress. As such, we have found that students gain a sense of satisfaction from working with this kind of task.

When I (Mike) taught High School English in West Africa, I reserved Fridays for our 'music day'. I introduced classic and modern American and British songs and, even without the use of audio recordings, I discovered that a few musically gifted students in every class would pick up on the tunes quickly and teach the class for me, often introducing imaginative variations and delightful choruses. I also discovered the motivating power of music: if the class began to fall behind the lesson plan, I only had to intone mildly, 'Maybe we won't have time for music this Friday . . .', and the class quickly buckled down.

Research links

Murphey, T. (1992) *Resource Books for Teachers: Music and Song*. Oxford: Oxford University Press.

Rosova, V. (2007) 'The use of music in teaching English: diploma thesis', Brno: Masaryk University, Faculty of Education, Department of English Language and Literature, 2007. http://is.muni.cz/th/84318/pedf_m/diploma_thesis_1.pdf.

Schoepp, K. (2001) 'Reasons for using songs in the ESL/EFL classroom', *Internet TESL Journal*, 7, pp. 1–4. http://iteslj.org/http://iteslj.org/Articles/Schoepp-Songs.html.

http://www.musicandlearning.com.

Activity: Finish the Story

Introduction

This activity is based on the instructor's ability in story-telling: setting the scene and creating suspense. The activity involves integrated skills work: intensive listening, writing and speaking. The activity also requires the students to use their creativity in providing a plausible ending for the story, thereby asking for a personal investment, a key element in building motivation. As a counterpoint to most listening activities that focus on getting the correct answer, this activity is open-ended. There are no correct answers and no set finishing point to the story, so the students can inject their own creativity into the task.

Aim
- listen to a narrative sequence
- apply creative faculties to extend the input
- develop these active listening strategies: comprehension monitoring, personal elaboration, creative elaboration, performance evaluation, repetition

Level Intermediate +

Time 20 minutes

Materials

A story that will interest your students and that is complex enough to make them think. Ideally, the story should include an intriguing set-up that lends itself to alternative continuations.

Preparation

Be ready to read or play the story and stop at the moment of greatest tension.

Procedure

1. (before listening) Explain to the students that you are going to tell them a story which they must finish in one sentence/three sentences/one paragraph/five minutes (whichever of these you choose). Explain that they will hear the story only once. Here is a sample script for the teacher:

> *'I'm going to tell you a story once and once only. Your task is to listen and write an ending. You will have exactly five minutes to write your ending. OK?'*

2. (listening) Read or tell the story. Students should not take notes.

3. (after listening) Get the students to retell the story in groups. They compare their endings. Volunteers may read out their endings to the class.

4. Tell the rest of the story *or* ask the students to decide on the best ending.

Variation

- Tell/play the story twice. The first time they listen, get the students to answer one or two gist questions.

- Increase the challenge by giving an exact word count: e.g. 'finish the story in exactly 100 words'.

Comments

The preliminary stage, during which we explain that the students will have to write the ending, is vital because it focuses the students' listening. As they listen, they are constantly considering and reconsidering possible endings. This exercises their critical faculties. The fact that they hear the story only once also motivates the students to listen attentively.

Research links

Erkaya, N. (2005) 'Benefits of using short stories in the EFL context', *Asian EFL Journal*, 8, pp. 1–13. http://www.asian-efl-journal.com/pta_nov_ore.pdf.

Wajnryb, R. (2003) 'Stories: Narrative activities for the language classroom', *Cambridge Handbooks for Language Teachers*. Cambridge: Cambridge University Press.

Sample Script

Audio File 4: Finish the Story

Please visit www.routledge.com/9781408296851 for an audio recording of this transcript.

Voyager 2 (say *hi* to ET!)

In 1977 we put a space probe called Voyager 2 into the solar system. The purpose was to make contact with aliens. In this space probe there was a record player and a record that was made especially for any extraterrestrials that may find our space probe.

This record contained numerous sounds that were supposed to represent humanity. The sounds included greetings in 55 human languages: '*Bonjour, buongiorno,* hello, *guten Tag*' etcetera, as well as 'hello' in whale language. It included the sounds of a baby crying, a couple kissing, and 90 minutes of music from all over the world – a Mexican mariachi, panpipes from Peru, an American Indian chant, a Japanese shakuhachi piece, Beethoven, Mozart, Louis Armstrong and Elvis Presley. It also had a message of peace read aloud by the Secretary General of the United Nations.

Anyway, I'd like you now to imagine that, one day, in a galaxy far, far away, an extraterrestrial finds Voyager 2. The extraterrestrial finds the record player and the record with the sounds and the message and the music, puts it on, and begins to listen. What happens next?

Ending 1 (creative): the aliens listen and decide to make contact with Earth, so they launch a message 10 million miles across the cosmos: 'Send more Elvis Presley'.

Ending 2 (factual): Funding for NASA's SETI (Search for Extraterrestrial Intelligence) programme was withdrawn in the early 1990s after three decades. It was decided that there were so many problems on Earth that there was not sufficient justification for the continued expenditure of taxpayers' money on the search.

Activity: Punchline

Introduction

Humour is a valuable resource for language learning. Some educators contend that the use of humour – and laughter – appeals to the kinaesthetic learning style. Although humour is a form of play, it serves several important 'serious' functions. Humour can help us relax, promote better social relationships, provide creative insights into our thinking processes, and improve retention of information. Including humour-inducing activities into our instruction can help break down affective barriers to learning. This activity is designed to make students laugh, contribute their own jokes, and examine the basis of humour. Because cultural humour is so often misunderstood, this activity can also provide fuel for productive discussions on cross-cultural misunderstandings.

Aim
- introduce humour into language lessons
- raise awareness of ways to deal with misunderstandings
- develop these active listening strategies: advance organising, comprehension monitoring, retrospective inferencing, mediating

Level Advanced

Time 30 minutes

Materials

A list of jokes divided into 'set up' and 'punchline'. See the worksheet for examples.

Possible sources:

http://iteslj.org/c/jokes.html

http://www.1000ventures.com/fun/fun_ps_j.html

Animal jokes http://animaljokes.resourcesforattorneys.com/index.php

History-related jokes http://history.inrebus.com/

Preparation

Prepare a worksheet that has the punchline of several jokes, but not the lead up. The worksheet should also have a rating space for 'funniness'.

Procedure

1. (before listening) Introduce the topic of the activity. Pass out the worksheet and read through Column 1: the 'punchlines'. Ask students if they can guess the first part of the joke.

2. Read the first part of each joke, and then have the students read the 'punchline' aloud. After the punchline, they are to rate their opinion of the joke.

   - Joke 1: What are two things that people never eat before breakfast?
   - Joke 2: Two women friends are having lunch together. One of them says to the other: 'Aren't you wearing your wedding ring on the wrong finger?'
   - Joke 3: My little brother often asks me to help him with his schoolwork. One day, he came home from school and showed me his geometry test. He said, 'Can you explain this to me? My teacher marked my answer wrong. The question was "Find x" and that's what I did.' Here's what he showed me . . .
   - Joke 4: Two cows are standing in a field. One of them says to the other, 'Are you worried about Mad Cow Disease?'
   - Joke 5: The English teacher says to her class, 'Today, we're going to talk about the tenses. Now, if I say "I am beautiful", which tense is it?'
   - Joke 6: Two ESL students went to Honolulu on holiday. Soon they began to argue about the correct way to pronounce the word 'Hawaii'. One student insisted that it's Hawaii, with a 'w' sound. The other student said it was pronounced like 'Havaii', with a 'v' sound. Finally, they saw an old native on the beach, and asked him which was correct. The old man said, 'It's "Havaii".' The student who was right was very happy, and thanked the old man.

3. (after listening) Ask students to decide what type of humour the joke is. See Part 2 of the worksheet.

4. Have students choose one joke they like. In pairs, they practise telling the joke, working particularly on rhythm and timing. (Optional: students choose a joke they like in their L1, and you work with them on rendering it into English.)

Variation

Joke competition. Have the class spend one minute thinking about their favourite joke. Tell your joke to the group (in English). Decide whose joke was funniest. Discuss in groups the humour (or lack of humour) in each joke, using the 'What's so funny?' list.

Comments

We have found that humour is a great ice-breaker for classes. Ongoing use of humour can create a relaxed learning environment, and the use of frequent jokes in the L2 can lead to improved listening. Telling jokes is only one form of humour. Other forms include exaggerating (for example, if you write nine-teeeeeeen on the board to indicate the appropriate stress), asking crazy questions (for example, 'Do you prefer Italian food or dog food?'), making intentional mistakes (for example, saying 'I'll return these assignments to you last Friday.').

Research links

Ahmed, A. 'Breaching cultural barriers with humor': http://www.youtube.com/watch?v=AexUlPFNcVw.

Sergio Aragones: Comic Artist http://www.youtube.com/watch?v=caHVIE6vItQ http://faculty.ksu.edu.sa/76518/Linguistics/books/translating%20jokes.pdf.

Wagner, M. and Urios-Aparisi, E. (2008) 'Pragmatics of humor in the foreign language classroom: learning (with) humor' in Putz, M. and Neff-Vanverselaer, J. (eds) *Developing Contrastive Pragmatics: Interlanguage and Cross-cultural Perspectives*. Berlin: DeGruyter.

Worksheet

Part 1. Listen to the first part of the joke. Read the completion aloud. Is it funny to you?

Joke	Not funny	A little funny	Very funny	I don't get it!
1. Lunch and dinner.				
2. The other one answers, 'Yes I am, I married the wrong man.'				
3. Here it is				
4. The other one says, 'No, it doesn't worry me. I'm a horse!'				
5. One student raises his hand and says, 'Obviously it's the past tense.'				
6. The man answered, 'You're velcome.'				

Photocopying of this worksheet is permitted © Taylor and Francis, an informa business, 2013

Part 2. What makes each joke 'funny'? Match each joke with a 'type' (there may be more than one answer).

This joke:

(a) is just silly (based on a ridiculous idea, circumstance or action)

(b) is a pun (uses words to create double meanings)

(c) is a riddle (requires figuring out something)

(d) is based on cultural stereotypes (exaggerates a simplified image of a group)

(e) gives you an unexpected answer (surprises you, makes you think about your logic)

(f) is about a misunderstanding (shows differences of language, culture or lifestyles)

(g) makes light of (or gives relief in) a difficult situation

(h) makes fun of (or insults) a naïve person

Frame 2
Top Down Frame

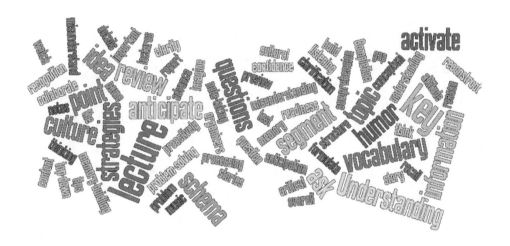

Introduction

The **Top Down Frame** deals with issues of attention, comprehension and memory. Top down listening is concerned with 'getting' what the speaker says: understanding big ideas, focusing on main points, recalling information, inferring ideas that are not clearly stated, synthesising and interpreting and organising what you understand.

The Top Down Frame addresses questions such as: How can I get my students to *focus on ideas* as they listen? Can I help my students to *think ahead*, predict more as they listen? How can I help them *keep going* even if they are having some problems? What can I do to develop *critical thinking*? How can I assist students to ask more *probing questions*? Can I help students *revise* their understanding after they listen? Is there something I can do to help students *recall* more of what they hear?

Proficiency in listening requires effective top down processing: focusing on main ideas, making predictions and drawing inferences, and recalling the organisation of what has been said. This frame presents activities that are proven to be useful in developing this key aspect of listening. For activities in this frame, it is helpful to have learners focus on developing strategies of focusing attention, inferencing and reviewing. (See Appendix 1 for a list of active listening strategies and examples of specific strategies the students can try out.)

Ten illustrative activities

The ten activities presented in the Top Down Frame are:

Activity	Description	Focuses on increasing attention span	Focuses on employing background knowledge	Focuses on pre-listening	Focuses on post-listening	Focuses on multi-modal processing
Guiding Objects	using images, objects and key words to build comprehension			■		■
Top 10 List	practising guided note-taking		■	■		■
Memories	reconstructing a narrative based on partial information			■	■	
KWL Chart	anticipating content using a chart		■	■		
Keep Doodling	creating a visual structure for a piece of listening input	■			■	
2-20-2 Pictures	using visuals to stimulate guesses about stories		■		■	
The Right Thing	using multiple perspectives to understand a story		■		■	
Good Question	using advance organiser questions to understand a lecture		■		■	
Split Notes	practising note-taking in an interactive manner			■		■
False Anecdote	listening for a lie in autobiographical stories		■			

(Darkened boxes indicate research links to the activities.)

Activity: Guiding Objects

Introduction

This activity works on increasing students' attention span by helping them activate expectations through imagining the significance and interrelationships of objects and images. As with most effective top down listening activities, this activity involves a pre-listening step and introduces some multi-modal processing (use of the 'guiding objects'.) Integrating verbal and visual information makes it easier to comprehend and reconstruct messages. The follow-up steps of the activity allow the students to create their own objects and stories. This type of appropriation activity allows the students to use the instructor's model and inject their own content.

Aim
- use images and objects to guide listening
- develop the habit of focusing on key information
- use key words to build comprehension and recall
- develop these active listening strategies: self-management, selective attention, predictive inferencing, visual elaboration

Level   Beginner +

Time   15 minutes

Materials

A question that you will answer in a short descriptive monologue.

Examples: What is your (father, mother, sister, brother) like? What is your (house, apartment, office, bedroom) like?

A few small objects that represent ideas to be included in your short talk.

Preparation

Choose one of the description questions. Select three or four objects you can bring to stimulate predictions about the topic. For example, if you are describing someone, the items could be: a copy of a magazine (she's a journalist), a spatula (she likes cooking), a shoelace (she enjoys hiking).

This is an example of what you might say, but it's best not to script it and instead just speak from key words, adding fillers to make it more like natural speech.

*'My sister Jenny is really **nice**. She lives in **Santa Rosa**. She's a **journalist** and has **two kids**. She's a **good cook** and loves going to **farmers' markets**. She also loves **reading** and **hiking**.'*

Procedure

1. (before listening) Write the question on the board: What is (your sister) like? Tell the students you're going to answer this question.

2. Hold up two or more objects that invoke images of the topic. Tell the students that your description will include ideas about these objects. Ask them to guess what the objects signify in this context. (Optional: pass the objects around the class to allow students to touch them and inspect them carefully.)

3. (listening) Begin your descriptive monologue. You don't need to use a script but be sure to include the key words you have prepared. (See the example.) The students should pay attention and not write.

4. (after listening) Ask the students to write down any key words they remember from your description.

5. Have the students work in pairs to reconstruct the description as completely as they can.

6. Tell the students to prepare a set of images (or objects if available) to answer the question: What is your (. . .) like? They should also prepare 10 or fewer key words.

7. The students work in pairs or small groups to deliver their descriptions to their partners. Partners can ask questions to get more information.

Variation

Key words only. Compose your own extended monologue and write down only the key words or phrases. Count up the words. Example question: What's my sister like?

Key words: nice, Santa Rosa, journalist, two kids, good cook, farmers' markets, reading, hiking.

*('My sister Jenny is really **nice**. She lives in **Santa Rosa**. She is a **journalist** and has **two kids**. She is a **good cook** and loves going to **farmer's markets**. She also loves **reading** and **hiking**.')*

Read your description to the class, explaining that students should write only key words. Be sure to use all of the key words. Score students based on how many key words they got (i.e. function words do not count).

Comments

We like this activity because it can be used with all levels of students and requires minimal preparation. It also presents a clear bridge from listening to speaking as it models for students how they can use just a few key words to prepare a longer turn when they are speaking. We also like it because it can be done frequently, with increasingly difficult input, as a refresher on using top down listening skills.

Research links

Bransford, J. (2003) *How People Learn: Brain, Mind, Experience, and School.* Washington, DC: National Academies.

Bransford, J. and Johnson, M. (2004) 'Contextual prerequisites for understanding: some investigations of comprehension and recall' in Balota, D. and Marsh, E. (eds) *Cognitive Psychology: Key Readings.* New York: Psychology Press.

Stimulating imagination and anticipation

a spatula   a magazine   shoelaces

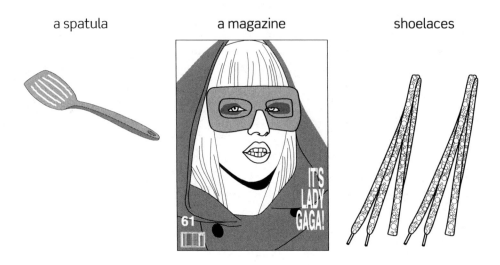

Example What's your sister like?

nice, Santa Rosa, journalist, two kids, good cook, farmers' markets, reading, hiking

Activity: Top 10 List

Introduction

Most students will be familiar with the idea of list-making in the context of shopping lists, 'top tips', lists of things to take on trips, media-generated 'Best of . . . (the year)' lists, etc. This activity uses the structure of a Top 10 list to serve as an **advance organiser**. Advance organisers enable students to pay attention longer because they help listeners to anticipate the next piece of information. The activity also encourages students to use their own background knowledge of a topic to guide their comprehension. Framing the activity as a Top 10 list also adds interest and piques the learners' curiosity.

Aim
- practise **top down processing** by using your own experiences
- practise **guided note-taking**
- develop these active listening strategies: persistent attention, world elaboration, predictive inferencing, noting, repetition

Level Beginner +

Time 20 minutes

Materials

An audio, video or text presentation that includes a salient list of items, steps in a process, or pieces of advice. Example:

Top 10 tips for making the most of your first week at college: http://www.youtube.com/watch?v=haf3EKZ_Flo (2 minutes).

Other ideas: Ten things employers want from a new hire: http://www.youtube.com/watch?v=XGdTTbXizYI.

Top 10 tips for better sleep: http://www.youtube.com/watch?v=qc5h0QXdZgo.

Top 10 tips for healthy cooking: http://www.youtube.com/watch?v=2RqJvHNzNOo.

Ten things you need to know about London: http://www.youtube.com/watch?v=16ztSXkbquo.

Preparation

1. Preview the content of the video or audio clip, or read through the text. Note key verb phrases that occur. Imagine yourself giving the advice, as you may need to paraphrase the content during the lesson.

2. Create a handout that guides listening to the list. (See sample worksheet.)

Procedure

1. (before listening) Preview the topic of the video or audio clip. Ask questions to activate interest.

2. Distribute your worksheet. Ask the students to anticipate any missing items on the worksheet. They can work in pairs to do this.

3. (listening) Narrate your list or play the video clip. Tell the students to note down completions for each tip.

4. When you have played the full audio once, let the students compare notes. If the students have missed any of the tips, play the audio one more time.

5. (after listening) Ask the students to identify which of the tips they consider most important.

6. Have the students work in pairs or small groups to compare their lists of 'top tips'. Students may wish to revise their own lists as they talk with classmates. Survey the class about which tips were considered the most important. Make a list on the board.

Variation

My top ten list. Have the students prepare their own lists of advice related to the topic of 'Tips for the First Week . . .' or on other topics of interest to them. These could include 'How to . . .' topics such as cooking a particular dish or learning English vocabulary. The students can rehearse this activity in pairs or groups, rotate to other groups to gain additional fluency practice, and then present their ideas to the class.

Comments

We think it's important in an activity like this to distinguish teaching from testing. Teaching involves testing, of course, but not all teaching needs to focus on testing. In this activity, it's not necessary to test whether the students have complete knowledge of what was said, or to check if they have taken complete notes. In the real world all listeners have to cope with partial understanding and, from this, make sense of what they understood.

Research links

Clement, J., Lennox, C., Frazier, L., Solórzano, H., Kisslinger, E., Beglar, D. and Murray, N. (2009) *Contemporary Topics*, 3rd edn. White Plains, NY: Pearson Education.

Kanaoka, Y. (2009) *Academic Listening Encounters: The Natural World.* Cambridge: Cambridge University Press.

Worksheet

Top 10 Tips for Getting the Most out of University in the First Week

1. Introduce _____.

2. Save _____.

3. Bring _____.

4. Remember: you may be _____,

   _____.

5. Make a list of _____

   so that you can plan _____.

6. Don't _____.
   You'll miss everything.

7. Personalise _____.

8. Make _____,

   so that you know _____ each week.

9. Choose _____.

10. Learn _____.

What are *your* top three tips from this list?

1. _____

2. _____

3. _____

What additional tips can you add?

- _____
- _____
- _____

Activity: Memories

Introduction

In this activity, the teacher or a fellow student tells a personal story involving memories. This kind of personalisation tends to promote greater affective involvement. The activity is structured as a jigsaw task, providing a different purpose for each listener, in order to teach the strategy of **selective listening**. As with all jigsaw tasks, it is important to have a clear pre-listening step so each student has a concrete focus as they listen. During the post-listening (reconstruction) phase of the activity, the students will be collaborating and helping each other to fill in details of the story. It then asks the students to construct similar memory narratives and share them with their classmates.

Aim • collaborate with others to build comprehension

  • develop these active listening strategies: selective attention, linguistic inferencing, personal elaboration, joint task construction

Level Beginner +

Time 30 minutes

Materials

None needed for Part 1. Prepare a task worksheet for Part 2. (See sample worksheet.)

Optional: some photos of your childhood, or general nostalgic photos about childhood.

Preparation

Think of a story you can tell from your childhood. Adjust the length and complexity to the students' level.

Procedure

Part I

1. (before listening) Tell the students you will give a short talk about a memory from your childhood. Write five to seven key words or phrases on the board as a kind of preview for the story. Check that the students understand this vocabulary before you begin.

2. Assign each student a task. See the sample script. Some students will have the same task.

3. (listening) Narrate your story. Encourage the students to take notes to complete their task as they listen.

4. (after listening) Put the students into groups of five. Each group should consist of one student who had each of the assigned tasks. Working together, they are to reconstruct the story as best they can.

Part II

1. (before listening) Ask students to prepare a short talk about one of their favourite childhood memories. They need to describe the event in as much detail as possible. Provide questions as prompts:

    How old were you? Do you remember what year this took place?
    Where was it?
    Who was with you?
    What happened?
    Why is this memory special?

2. Give the students several minutes to think, make notes and ask about any vocabulary they may need for their presentations.

3. Allow the students to rehearse their stories in pairs.

4. Before the mini-presentations, assign each student one listening task only. (Some students will have the same task, but assign the tasks so that approximately the same number of students have the same task.) You can assign tasks by having index cards with one task written on each, or simply present all five tasks, and have the students count off to have their task assigned.

    Task 1: Note the age of the speaker when the memory took place.

    Task 2: Note the place where each memory took place.

    Task 3: Note the people involved in each memory.

    Task 4: Note what happened: list one or two key actions.

    Task 5: Note why this is a special memory.

5. (listening) Have the students give their mini-talks. Set a time limit for each student presentation. When they are not speaking, students must be listening and taking notes according to their assigned tasks.

6. (after listening) Ask the students their opinion of the activity. Did having small tasks help them pay attention? Did it help them understand more? Do they want to do this kind of activity again? How would they improve the activity?

Variation

Group story telling. Following the teacher's model story, have the students compose their own story – using key words only. Then the students work in groups, telling their stories. As a follow-up assignment, students write their story.

Comments

We like this kind of activity because it uses the 'Teaching Unplugged' approach, taking the learner as the starting point for all language work, drawing upon each learner's experiences, thoughts, feelings and opinions. In this approach, listening activities become a means of exploring and articulating the students' own ideas.

Research links

Derwing, T. (1996) 'Elaborative detail: help or hindrance to the NNS listener?' *Studies in Second Language Acquisition*, 18, pp. 283–297.

Goh, C. (2000) 'A cognitive perspective on language learners' listening comprehension', *System*, 28, pp. 55–75.

Griffiths, G. and Keohane, K. (2000) *Personalising Language Learning*. Cambridge: Cambridge University Press.

Hasan, A. (2000) 'Learners' perceptions of listening comprehension problems', *Language, Culture and Curriculum*, 13, pp. 137–153.

Jensen, E. and Vinther, T. (2003) 'Exact repetition as input enhancement in second language acquisition', *Language Learning*, 53, pp. 373–428.

Sample Script

Intermediate level

Audio File 5: Memories

Please visit www.routledge.com/9781408296851 for an audio recording of this transcript.

This is something that happened when I was in first grade, I guess when I was six or seven years old. We had a very mean teacher, Mrs Martin, and I remember that I was terrified of her when she got angry. One day she told us during the morning recess that we were to be completely silent during the break, and she assigned a monitor to be sure that we didn't talk. I mean, that's kind of ridiculous, right, to expect that a bunch of 7-year-olds can be quiet for 15 minutes, but that's what she said. So, during the break, this monitor, his name was Steve, who was one of my best friends, and I guess he was kind of the teacher's pet, too and, um, when we got back to class, Mrs Martin asks Steve, 'Did anyone talk during the break?' And Steve says, 'Only one person.' And then he says my name! I practically died. I just couldn't believe it. I mean, maybe I talked a little, but not any more than anybody else. Betrayed by my friend. So Mrs Martin made me stay in the class during lunch time and I had to write, 'I will not talk during recess' like a hundred times. So the whole class goes out to lunch, and I have to stay there all alone, and I had tears streaming down my face, and I start writing, like one or two times, 'I will not talk' and then I think, 'I'm not doing this. It's not fair.' And I get up and leave the school and start walking home. I mean, I wasn't even sure where my home was because I took the bus to school, but I start off in the general direction. And after about 20 minutes, a big yellow school bus pulls up, the door cranks open, and the driver asks me to get on. And I look at him and think, like, no way, I'm not going back to that school, ever, and I say, 'No, thanks.' But he gets off the bus and comes and puts an arm around my shoulder, and says, 'I'll take you home.' And I finally I get on the bus and he takes me home. And, man, long story short, I got in a world of trouble with my parents and the teacher and the principal. But I didn't care. I thought, if something's not fair, you have to take a stand. You can't just put up with it.

Worksheet

Photo courtesy of Veer/Corbis Images

Task 1: Note *the age or the year* the memories took place.

Task 2: Note *the place* where each memory took place.

Task 3: Note *the people* involved in each memory.

Task 4: Note *what happened*: list one or two key actions.

Task 5: Note *why this is a special memory*.

Name of Presenter

Your task

Other information for summary

Activity: KWL Chart

Introduction

This activity features an advance organiser, a structure by which the students can use background knowledge to make predictions about what they will hear. In effect, the students are setting their own pre-listening comprehension questions by asking themselves: Are my assumptions about the topic accurate? Will my questions be answered? Developing this prediction strategy is an important confidence builder for students. Rather than a third party (such as the teacher or a materials writer) anticipating what students know and setting comprehension questions, the students themselves create a focus for listening. Because students have invested in the topic by saying what they know and by developing questions, they are more likely to listen actively.

Aim   • activate prior knowledge of a given topic
- use personalised questions as a framework for listening
- develop these active listening strategies: advance organising, predictive inferencing, world elaboration, revision evaluation, summarisation

Level   Beginner +

Time   20 minutes

Materials

A factual recording or text to be read aloud and a model of a KWL chart (see the sample) or a photocopied handout of the chart.

Possible sources:

- Course book recordings
- For short 'bites' of information that can be read aloud, as well as longer essays, on most topics, see http://en.wikipedia.org.

Preparation

Listen to the factual recording. Have an example of a KWL chart ready or be prepared to draw a model of one on the board.

Procedure

1. (before listening) Tell the students the topic and show them a KWL chart (see Chart 1 on the sample). Explain that the K stands for know. In this

section, they write what they know about the topic. The W stands for want to know. Here they formulate genuine questions about the topic. The L stands for learned. They leave this column blank until after hearing the recording.

2. Ask students to fill in the first two columns (see Chart 2 on the sample).

3. After a few minutes, put the students in pairs to compare what they wrote.

4. As a group, the students volunteer some of their suggestions for the first two columns.

5. (listening) Explain that they will listen to confirm their ideas from the Know column, and to find out if any of their questions in the Want to know column are answered. Play the recording.

6. (after listening) The first feedback stage involves discussing which of the students' ideas have been mentioned and which of their questions have been answered.

7. After a second listening, the students write in the third column everything new that they learned. Again, this is followed by a feedback stage.

8. (Optional) As a follow-up step, the students go away and find out answers to any of their questions that were not answered.

Variation

KNL chart. In this variation, K stands for know, N stands for need to learn, and L stands for learned. This variation is commonly used in content-based classes at secondary school level.

Comments

We have used KWL charts for all types of student. While the charts are commonly used in teaching children school subjects such as History and Geography, they are just as appropriate in ESL/EFL contexts.

Research links

http://www.projectlearnet.org/tutorials/advance_organizers.html.
http://www.glnd.k12.va.us/resources/graphicalorganizers/.
http://www.education.com/reference/article/K-W-L-charts-classroom/.

Sample Charts

KWL Chart 1

K: What I know	W: What I want to know	L: What I learned

KWL Chart 2 – Topic: The Olympic Games

K	W	L
Athletic competition Began in Greece Happens every four years A different country hosts it every four years	When did it start? How many countries take part? How many sports are involved?	

KWL Chart 3 – Topic: The Olympic Games

K	W	L
Athletic competition Began in Greece Happens every four years A different country hosts it every four years	When did it start? How many countries take part? How many sports are involved?	*Started: eighth century BC and again in 1894* *200 countries* *25–30 sports* *Modern Games were begun by a Frenchman* *Second biggest sports event* *Winter Games and Paralympic Games (disabled athletes)*

Sample Script

The Ancient Olympic Games can be traced back to Greece in the eighth century BC, but the modern version began when a Frenchman, Pierre de Coubertin, founded the International Olympic Committee in 1894. The IOC, as it is commonly known, became the governing body of the Olympic Games.

After the Football World Cup, the Olympics is the world's second biggest sporting event. Taking place every four years in a different city, the Olympics involves over 10,000 athletes from over 200 countries. There are around 25–30 different sports played, though this changes as sports are added and subtracted from the list.

Besides the main Olympics, there are now offshoots, which include the Winter Games and the Paralympic Games for athletes with disabilities.

Activity: Keep Doodling

Introduction

The idea of this activity is for students to make a visual representation of the listening passage. All good texts prompt listeners or readers to create images in their minds. Readers often describe a fine passage of writing as if they were there, and good storytellers use vivid phrasing so that listeners are momentarily transported in time and place. This activity uses transposition – shifting the mode from the aural to the visual – as a way to get students interacting more deeply with the passage and interpreting it in a personal way to create something new. This interaction with the input is one of the hallmarks of active listening.

Aim
- listen for gist and detail
- develop a visual structure for a passage
- develop these active listening strategies: visual elaboration, multi-modal inference

Level Advanced

Time 25 minutes

Materials

Go to youtube.com and type in 'RSA Animate'. Choose a few minutes of any suitable lecture, e.g. the first two minutes of Steven Pinker: Language as a window into human nature http://www.youtube.com/watch?v=3-son3EJTrU

or Philip Zimbardo: The secret powers of time http://www.youtube.com/watch?v=A3oliH7BLmg

Note: these lectures are authentic and contain some difficult concepts. They are suitable only for genuinely advanced students.

Alternatively, choose an extended listening passage – it could be a lecture or story – of up to five minutes, e.g. from http://storycorps.org/listen/

Preparation

1. Prepare to show the clip with RSA Animation. Include a few questions (a) about the content of the clip, and (b) about the animation: what the

students think of it, how it enhanced or detracted from the message, why it might be useful to visually represent a listening passage.

2. Write some straightforward gist questions for the extended passage if the passage is not from ELT materials.

Procedure

1. (before listening) Do your pre-listening work, naming the topic, activating schemata and giving a reason for listening. This may include some straightforward gist questions.

2. Play the clip.

3. Give students a few minutes to compare their answers.

4. Elicit feedback from the whole group and ask the students about the effect of the animation: whether the students like it or not, whether it helped them to understand the passage, why drawing might be a useful while-listening or post-listening technique.

5. Tell the students they will listen to a passage that is separate but related (in theme) twice. The first time they should listen for meaning only and answer one or two gist questions.

6. (listening) The students listen and answer the question(s) in groups of three or four.

7. Explain that they will listen again and will afterwards need to make a visual representation of the text in groups. Say that this need not only be drawings; it could be graphs, charts, diagrams, doodles.

8. Play/read the passage again. Give the students several minutes to depict the passage visually on large pieces of paper.

9. (after listening) Ask the students to work with people from other groups. The students explain to one another how and why their group chose to depict the passage the way they did.

Variation

- Provide the students with a framework for their visual representation (see the 'ideas web' sample).

- Use only the RSA animation sequence as your input. Then get students to deliver short, related talks in groups. The other students work collaboratively to produce visual representations of their classmates' talks.

Comments

Because many students lack confidence in their ability to draw, we encourage them to incorporate both words and illustrations in their representations, as in the RSA animations. (The RSA Animate series was conceived as an innovative way of making the Royal Society of the Arts public events programme more accessible to viewers.) Simple acts, such as writing in non-linear sequence or incorporating a graphic here and there, can be very useful for alerting students to other modes of representation besides linear notes. The key motivational element in the activity is for students to get excited enough about the input that they keep doodling and keep trusting their impressions and interpretations of the text.

Research links

Baddeley, A.D. (2007) *Working Memory, Thought and Action*. Oxford: Oxford University Press.

Ruhe, V. (1996) 'Graphics and listening comprehension', *TESL Canada Journal*, 14, pp. 45–59. Retrieved from: http://journals.sfu.ca/tesl/index.php/tesl/article/viewFile/677/508.

Sample Visual

Idea web

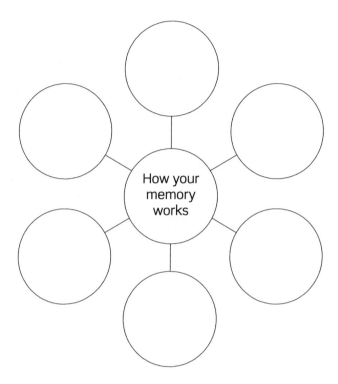

Activity: 2-20-2 Pictures

Introduction

In this activity, before listening, the students activate information about the topic in multiple ways: guessing, describing and speculating. While speculating, they draw upon all of their knowledge of the world, which is the basis of top down processing. Top down processing is always assisted by some explicit form of pre-listening activity like this. During this pre-listening stage, the students will be rehearsing various vocabulary items, saying the words out loud several times. This is a useful pre-listening step that enables quicker recognition of vocabulary during the listening phase of the activity.

Aim	• practise speculating about pictures before listening
	• listen for confirmation of ideas
	• develop these active listening strategies: world elaboration, visual elaboration, multimodal inferencing, retrospective inferencing, repetition
Level	Intermediate +
Time	15 minutes

Materials

A short news extract (less than 150 words, about 1–2 minutes when narrated) that has a surprise ending (see sample script), and two photos or images that suggest the theme of the story, but don't give away the ending (see sample visual). The images may be on pieces of paper or presented as PowerPoint slides.

Possible sources:

http://www.google.com/imghp
http://www.freeimages.co.uk/index.htm
http://www.sxc.hu/

Preparation

Prepare the two pictures that relate to the recording. Make sure that all the students will be able to see them clearly.

Procedure

1. (before listening) Show the pictures for two seconds each. In pairs, the students say everything they remember about them – objects, weather conditions, number of people, etc. – and try to compare and contrast the pictures. This phase may be very short.

2. Show the pictures for 20 seconds. The students repeat stage 1 but in more detail.

3. Show the pictures for two minutes. The students describe the pictures, saying what the theme is, how the pictures are connected, what's happening now, what might have happened before the photos were taken, and what might happen next.

4. (listening) The students listen to the recording to check their ideas.

Comments

We have found that, because the students invest something of themselves – their guesses, ideas and background knowledge – before listening, they are more likely to listen actively to confirm their speculations.

Research links

Goldstein, B. (2009) *Working with Images.* Cambridge: Cambridge University Press.

Keddie, J. (2009) *Images – Resource Books for Teachers.* Oxford: Oxford University Press.

Sample Visual

Low intermediate

Picture 1   Picture 2
Photos courtesy of Veer/Corbis Images

Audio File 6: 2-20-2 Pictures

Please visit www.routledge.com/9781408296851 for an audio recording of this transcript.

> A car thief started a 6-year prison sentence after stealing 36 cars in order to clean them. All the cars were stolen from car showrooms. The court was told that David Blain, a cleaner who doesn't own a car, walked into car showrooms and asked to test-drive a car. He then drove from the showroom and didn't return. Every car was later found at the side of the road, absolutely spotless inside and out. Blain washed and cleaned each one before leaving it. He was once called 'the man you would most want to steal your car' by one judge. Blain's wife, Mary, a 48-year-old nurse, said after the case that their marriage was over. She told reporters, 'He looked after the cars better than me.'

Activity: The Right Thing

Introduction

Problem stories, in which characters struggle to make a difficult or controversial decision, can be very effective in engaging students, particularly if they have some background knowledge concerning the topic of the controversy. This activity presents a common dilemma, and asks students to try to understand it from each character's point of view. At the same time, the student's own response will inevitably depend on their background – the cultural and familial values they bring to the class. This activity asks for personal investment – the listener has to become curious about the content and concerned with the outcome.

Aim
- practise critical thinking
- develop comprehension of multiple perspectives
- develop these active listening strategies: personal elaboration, retrospective inferencing, visual elaboration, problem evaluation, performance evaluation

Level Intermediate +

Time 20 minutes

Materials

Reading, listening or viewing material that presents a controversy – at least two sides of a problem.

Possible sources:

- Summaries of legal cases involving disputes. For example, Attorney George Boyle has prepared a series of scenarios, You Be the Judge: http://www.georgeboyle.com/judge.html, which presents actual cases that judges have had to arbitrate.

- Advice columns that present everyday problems, often involving relationships. For example, Ask Jaqi http://www.askjaqi.com/ provides advice for teenagers from a teen's perspective.

- TV court cases, for example: *Judy Judge* episodes available on YouTube: Spoiled brat http://www.youtube.com/watch?v=deYSnFYGSCO.

- A variety of social issues, involving relationships and moral dilemmas. Impact Issues series http://www.impactseries.com.

Preparation

Assign two students the roles of Toshi and Susan. Have them read over the script and prepare to perform it in front of the class. (Option: pre-record the script with yourself and a colleague.)

Procedure

1. (before listening) Present the picture of Toshi's mother. Give a brief background of the situation: Toshi and his wife Susan live in Tokyo, with their two young daughters. They're talking about Toshi's mother. They have to make a decision. What do you think the decision is?

2. Tell the listeners their task is to try to understand the problem from Toshi's perspective, from Susan's perspective, from their daughters' perspective, and from Toshi's mother's perspective. Explain that perspective = point of view, what's important to them, what's the right thing to do in their opinion.

3. (listening) Have the two assigned students read the dialogue on the worksheet or play a pre-recorded version of the dialogue (use audio file or make your own recording).

4. (after listening) In small groups, the students attempt to outline the perspective of each of the four parties involved:

   Toshi

   Susan

   Mother

   Daughters

6. After about five minutes of small group discussion, poll the class: What is the 'right thing' to do in this situation?

7. As a follow up, have the students think of similar dilemmas they have faced in their lives, or dilemmas of people they know. They can write a short dialogue to show two different points of view concerning the dilemma. They can read their dialogues and ask the class for a similar type of response: What is the perspective of each person? What's the right thing to do?

Variation

Jigsaw listening. Prepare two monologues about the same story or case, each told from a contrasting point of view. Divide the students into groups:

A students hear (or read) one version; B students hear the other version. The students then pair up in A–B dyads to exchange the differing perspectives, before they decide 'the right thing' to do.

Comments

Early in our careers, we utilised court cases as the basis for jigsaw-type lessons. We were initially surprised at the interest level of our students in the often mundane cases, such as neighbour disputes over who owns the apples that fall from a tree growing on their property line. Perhaps it was the everyday nature of the issues that allowed students to relate to the stories and readily express their opinions.

Research links

Berne, J. (1995) 'How does varying pre-listening activities affect second language listening comprehension?' *Hispania*, 78, pp. 316–329.

Berne, J. (2004) 'Listening comprehension strategies: a review of the literature', *Foreign Language Annals*, 37, pp. 521–533.

Chang, A. (2007) 'The impact of vocabulary preparation on listening comprehension, confidence and strategy use', *System*, 35, pp. 534–550.

Worksheet

Using 'problem stories'

Photo courtesy of Getty Images

Audio File 7: The Right Thing

Please visit www.routledge.com/9781408296851 for an audio recording of this transcript.

Toshi:	We have to face the facts. Mother is too old to live alone.
Susan:	Yes, you're probably right. She is getting older and, since your father died, she seems very forgetful.
Toshi:	She'll need someone to take care of her in the near future.
Susan:	Maybe it's time to consider a nursing home.
Toshi:	No, no way. It's way too soon to think about that.
Susan:	Then maybe we could hire a nurse. You know, someone who could come in during the day, clean the house, cook meals, make sure she takes her pills.

Toshi: Well, you know nurses don't do all of those things. You're thinking of a nurse *and* a housekeeper. Who would pay for that?

Susan: Well, Toshi, I don't know what other choice we have.

Toshi: You probably know what I'm going to say. I'm the only child. And in my culture, that means we have to take care of her. She can come to live with us!

Susan: Wait a minute! Here? In this tiny place? There are only two bedrooms. We can't move our two daughters out of their small bedroom.

Toshi: What else can we do? We can't afford to hire a nurse and a housekeeper. I'm her only child. The rest of her family have all passed away. We have to take care of her.

Susan: Maybe we could look for a bigger house.

Toshi: You know that's impossible. We tried looking last year and everything in this area is way over our budget.

Susan: Then let's move away from the city. Someplace cheaper.

Toshi: But then I would be too far away from my office. You know how much I hate long commutes. And we'd have to move the girls to new schools. No, the only solution is for Mother to live with us here.

Activity: Good Question!

Introduction

Active listening requires thinking ahead to anticipate the type of information you will need to process. This activity uses different types of questions to help focus attention. The questions, if used as a pre-listening prompt rather than as a comprehension test, help the listener to *anticipate* what is coming in the next extract. This form of presentation assists short-term memory in processing new information. The goal is for students to become active listeners, and begin to *pose their own questions* as they listen. This activity models the range of inductive and deductive questions that a listener can pose.

Aim
- increase attention during listening
- practise asking different types of questions
- develop these active listening strategies: question elaboration, double-check monitoring, performance evaluation, revision evaluation, resourcing

Level Intermediate +

Time 20 minutes

Materials

Video or audio extracts up to 10 minutes in length

Possible sources:

(9:30) Video: History of Leonardo da Vinci: http://www.youtube.com/watch?v=Y_3qOFuheB4

(4:10) Video: Short biography of Marilyn Monroe (Norma Jean King): http://www.biography.com/people/marilyn-monroe-9412123/videos/marilyn-monroe-mini-bio-2078946296

(1:30) Audio: Short biography of Amy Winehouse: http://www.famouspeople lessons.com/a/amy_winehouse.mp3

Preparation

1. Select an extract and divide it into four or five shorter segments, preferably ones that have subtitles or natural pauses. For example, in the worksheet we divided the Leonardo da Vinci extract as follows:

History of Leonardo da Vinci
0:00–1:10 Introduction
1:10–3:00 Historical Context
3:00–5:45 The Young Leonardo
5:45–8:10 Leonardo's Teens
8:10–9:30 Leonardo's Challenge

2. For each segment, prepare two to four questions that you can pose *before* the students listen to that segment. Don't worry about asking questions to test exhaustive understanding. Instead, ask questions that rouse curiosity. Consider posing questions from each of these categories:

- Understanding the *gist* (main ideas and supporting information)
- Understanding details (minor facts and fine points)
- Understanding attitude (the speaker's attitudes and opinions about the topic)
- Deep understanding (implications that go beyond what was said)

Procedure

1. (before listening) Introduce the extract by showing images of the subject of the bio you've chosen. Ask the students what they know about this person. Write some key words on the board.

2. Distribute the worksheets. Tell the students the biography will be in short segments. Dictate some questions to be answered by the students in pairs after listening. (Note: the worksheet has the questions already written. Delete these before duplicating it or, if you wish to skip the dictation step, distribute as is.)

3. (listening) Play the video. At the end of each segment, pause the video. Give the students a few minutes to discuss the questions and note down their answers.

4. After they answer the questions, have the students write their own questions (things they would like to know) and any comments they may have about the segment.

5. Go over the answers to each segment before starting the next segment. Have students pose their questions. You need not know the answers; you can simply say, 'That's a good question.'

6. After playing the video in segments, play it again in its entirety.

7. (after listening) Assign a follow-up activity: writing a short summary of the talk, making a list of new vocabulary, or an out-of-class research assignment that extends the topic of the video.

Variation

Screen captures. After the students have watched the biography (or any video segment), show several screen captures. Ask the students what they can recall of the scene you are showing. Who is in the scene? What is in the scene? What are the characters talking about? Is this scene important to the story? How so?

Comments

We often use collaborative techniques in classes, particularly in content-oriented classes. The stronger listeners generally do not mind helping out the weaker listeners, and the weaker listeners can pick up valuable strategies from their classmates. By having learners collaborate during the recall segments of the lesson, we are showing that listening can be a collaborative activity.

Research links

Graesser, A. and Person, N. (1994) 'Question asking during tutoring', *American Educational Research Journal*, 31, pp. 104–137.

Lustig, C., May, C. and Hasher, L. (2001) 'Working memory span and the role of proactive interference', *Journal of Experimental Psychology*, 130, pp. 199–207.

Worksheet

Part I

0:00–1:10 Introduction
Question 1: What drove Leonardo to succeed?

Your answer	Question or comment?

Question 2: The narrator says Leonardo grew up 'in a place where life was cheap'. What does he mean by this?

Your answer	Question or comment?

Part II

1:10–3:00 Historical Context
Question 1: Why was the fall of Constantinople (1453) significant in Leonardo's life?

Your answer	Question or comment?

Question 2: Why is the interviewee (Mr DeVries) excited about humanism?

Your answer	Question or comment?

Part III

3:00–5:45 The Young Leonardo
Question 1: What is an example of something that came out of the Renaissance?

Your answer	Question or comment?

Question 2: Describe the circumstances in which Leonardo grew up.

Your answer	Question or comment?

Part IV

5:45–8:10 Leonardo's Teens
Question 1: The narrator suggests that Leonardo developed 'a code'. What is your interpretation of this?

Your answer	Question or comment?

Question 2: What was the relationship of the Medici family to Leonardo?

Your answer	Question or comment?

Part V

8:10–9:30 Leonardo's Challenge
Question 1: How significant are the guilds to Leonardo?

Your answer	Question or comment?

Question 2: What do you think the next segment of the biography will cover?

Your answer	Question or comment?

Activity: Split Notes

Introduction

This activity provides instruction in how to use notes to review and reconstruct important information. Because note-taking is a type of listening activity that has strong face validity, most learners will be willing to learn new techniques for enhancing their note-taking and recall skills. As with other top-down listening activities, both the pre-listening preparation and planning step and the post-listening review step are vital. Indeed, note-taking is most effective when it is used for active review and reconstruction of information *after* the lecture is over.

Aim • try out a practical method of taking and using notes

  • focus on key words

  • develop these active listening strategies: self-management, repetition, noting, summarisation

Level   Intermediate +

Time   30 minutes

Materials

A lecture text or series of texts that you can divide into three-minute segments. (See the sample worksheet.)

You can also introduce some formal systems of note-taking, such as:

the Cornell System, from James Madison University:
http://coe.jmu.edu/LearningToolbox/cornellnotes.html;

the Outlining Method, from Pedagog Education
http://www.study-skills.ca/blog/2008/03/04/taking-notes-outline-method/; or

the Sentence Method, from Cal Poly University
http://sas.calpoly.edu/asc/ssl/notetakingsystems.html#sentence

Preparation

Read over the lecture script that you will deliver. Underline key words that you wish to emphasise and you intend for students to note down. Mentally rehearse how you will deliver the lecture, segment by segment, including any paraphrases you may give.

Procedure

1. (before listening) Tell the students the topic of the lecture: Addiction. Ask them to brainstorm any ideas and any key words they think they might hear. Or give guided questions: What is addiction? What kinds of things are people addicted to? How does addiction develop? How can it be cured?

2. Pass out the worksheets (incomplete notes), with the key words in the lecture in the right column. Go over any difficult vocabulary.

3. (listening) Tell the students you are going to practise a note-taking technique called 'Split Notes'. For this activity, they will already have key words from the lecture written down for them. (When you repeat this type of activity, they will need to write the key words themselves.) Ask them to follow the lecture as you give it, pointing to the key words and phrases as you mention them.

4. (after listening) For each key word, the students, in pairs, compose a question for which the key word is the answer. (See the sample.) Do the first few together as a class.

5. (Optional) Using the worksheet, have the students discuss which of the ideas are main points and which are additional detail.

Variation

Guided note-taking. Preview the topic of the lecture. Ask questions to stimulate ideas that will come up in the lecture. When you deliver the lecture, stop after you have said a key word or phrase. The students are to include the word or phrase you have just uttered in their notes. After the lecture, students go back and make sentences using the key words.

Comments

We sometimes use the '5 R Method'. The 5 Rs are: Record (use headings and identify main points), Reduce (revise notes and *condense to key information* only within 24 hours of the lecture), Recite (talk *aloud* – to yourself or a study partner – saying the ideas *in your own words*), Reflect (*think about* the key points – and if possible *read more about* or write something related to these points); Review (*compose and answer questions* related to the ideas in your notes). It's not essential to follow all of these Rs all of the time, but it is important to keep the principles active during the study process.

Research links

Armbruster, B.B. (2000) 'Taking notes from lectures' in Flippo, R. and Caverly, D. (eds), *Handbook of College Reading and Study Strategy Research*, pp. 175–199. Mahwah, NJ: Erlbaum.

Carrell, P., Dunkel, P. and Mollaun, P. (2002) *The Effects of Notetaking, Lecture Length and Topic on the Listening Component of TOEFL*. Princeton, NJ: Educational Testing Service.

Cornelius, T. and Owen-De Schryver, J. (2008) 'Differential effects of full and partial notes on learning outcomes and attendance', *Teaching of Psychology*, 35, pp. 6–11.

Murphy, T.M. and Cross, V. (2002) 'Should students get the instructor's lecture notes?', *Journal of Biological Education*, 36, pp. 72–75.

Neef, N.A., McCord, B.E., and Ferreri, S.J. (2006) 'Effects of guided notes versus completed notes during lectures on college students' quiz performance', *Journal of Applied Behavior Analysis*, 39, pp. 123–130.

Sample Script

Lecture

Audio File 8: Split Notes

Please visit www.routledge.com/9781408296851 for an audio recording of this transcript.

Today, our lecture is about *addiction*. Most people think addiction means an *uncontrollable habit*, for example an addiction to using a *substance*, a substance such as alcohol or other drugs. These *substances* can be *addictive* because for some people – for many people – they have *physical effects* on the body and, particularly they have effects on the *brain*, so people who *regularly use* these substances, and regularly use them in *large amounts*, will tend to become addicted.

More recently, however, we have come to realise that people can also develop *addictions to behaviours*. OK? So we have addictions not just to substances, but also to behaviours, such as *gambling*. And while gambling may be considered a *risky activity*, we also see addictions to quite *ordinary* – and *necessary* – *activities* such as exercise and eating. These everyday activities can become addictive. Right? What all these activities have in common is that the person doing them finds *relief* in them in some way. This means that the substance or behaviour seems to *solve* some *problem* for them.

Worksheet

Incomplete version of note-taking sheet

Questions	Key words and phrases
	addiction
	uncontrollable habit
	substance
	addictive
	physical effects
	brain
	regularly use
	large amounts
	addictions to behaviours
	gambling
	risky activity
	necessary activities
	relief
	solve a problem

Sample

Completed version of note-taking sheet

Questions	Key words and phrases
What is the topic of the lecture?	addiction
What is an addiction?	uncontrollable habit
What do we get addicted to?	substance
What is the adjective of *addiction*?	addictive
What is a result of addiction?	physical effects
Where do we feel addiction effects?	brain
How do you get addicted?	regularly use
Do you need small or large amounts?	large amounts
What else is addiction?	addictions to behaviours
What is an example of addictive behaviour?	gambling
Is gambling ordinary activity?	risky activity
What other behaviour is addictive?	necessary activities
Why do we repeat addictive behaviours?	relief
What is the purpose of addiction?	solve a problem

Activity: False Anecdote

Introduction

This activity features extended speaking as well as extended listening. The students listen to find out which among several anecdotes is untrue – a task that requires them to use their background knowledge of the speaker and that tends to increase attention span. The activity is enjoyable for several reasons: the personalised aspect of the anecdotes, the element of sleuthing required to work out which anecdote is invented and, from the speaker's point of view, it is fun to try to deceive people in a game-like situation. This tends to motivate the speakers to become creative, inventing non-existent details in order to sound plausible.

Aim	• practise intensive listening
	• listen critically (and ask follow-up questions)
	• develop these active listening strategies: directed attention, world elaboration, joint task construction
Level	Intermediate +
Time	45 minutes

Materials

Home-made poster of five of the teacher's experiences.

Preparation

1. Make a poster (see sample visual) of five interesting experiences you have had. One must be a complete invention, though plausible. The poster should consist of five simple line drawings or illustrations of the experiences.

2. Prepare to describe the experiences to the class, grading your language as appropriate.

Procedure

1. (before listening) Stick your poster on the board.

2. Tell the class to approach the board so they can see the pictures. Explain that they will listen to you describing five stories about yourself but one never happened. Their listening task is to guess which one is not true.

3. (listening) Tell your stories. Speak at natural speed but leave longer pauses and grade your vocabulary to the students' level.

4. Offer to answer any questions about the stories. At this stage, the listeners become interrogators and try to catch out the speaker.

5. The students guess which one is not true. Tell them the answer.

6. (after listening) Hand out poster paper (see sample visual) and explain that the students must now do the same as you. Give them plenty of time for their illustrations and time to plan what they will say.

7. Put the students into groups of three or four. They put up their posters around the walls and take turns to do stages 3–5.

8. Change the groups so the students are working with new partners and have them repeat stage 7.

Variations

- For large classes you may need to draw big illustrations on the board so everyone can see. For all but the most artistically talented teachers, we suggest doing this before the class starts!

- Add a writing stage: the students choose one of their true stories and write the anecdote for homework.

Comments

In our classes, students tend to find this activity highly motivating. In terms of listening, their interest is first piqued by the drawings. Then it is the personal nature of the anecdotes (and the pleasure in guessing which is not true) that captures their attention. They genuinely wish to listen in order to both 'get' the story and to uncover the invention!

Research links

Two experienced teachers talk about using anecdotes:

http://www.youtube.com/watch?v=CRnBC7F_WB0.

Sample Visual

Story 1:

The first time I . . .

Story 2:

The time I almost . . .

Story 3:

How I became . . .

Story 4:

The most amazing . . .

Story 5:

The one thing I . . .

Frame 3
Bottom Up Frame

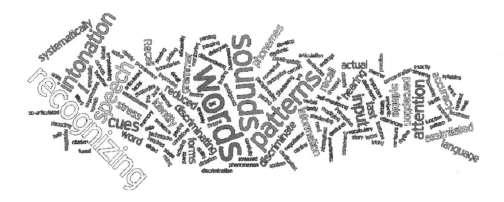

Introduction

The **Bottom Up Frame** deals with perception and decoding of spoken language. Perception refers to hearing the actual sounds and rhythmic patterns that the speaker articulates. Decoding refers to recognising words and parsing the grammar of what you hear *in real time*. All of these interdependent processes require speed and accuracy, and an ability to infer (in just a millisecond) what was missed. The point of activities in this frame is to strengthen perception so that much of bottom up processing can happen automatically.

This frame addresses questions such as: How much are my students actually 'getting' from the input, and how much are they 'missing'? How can I get my students to attend more closely to what the speaker says? How can I get students to become more curious about the actual sounds they are hearing? Can I help my students guess the parts of the input they are *not* hearing clearly? What can I do to get students to recognise vocabulary even from reduced and assimilated sounds? How can I get my students to process the grammar of the input? Is there something I can do to counter the influences of the students' L1 on their perception of the L2?

Bottom up listening skill is essential for achieving advanced proficiency in the L2. This frame presents activities that are proven to be useful in developing this aspect of proficiency, for learners of all levels. With activities in this frame, it is helpful to have learners focus on developing strategies of focusing attention, monitoring and reviewing. (See Appendix 1 for a list of active listening strategies and examples of specific strategies the students can try out.)

Ten illustrative activities

The ten activities presented in the Bottom Up Frame are:

Activity	Description	Focus on learning L2 phonology	Focus on adjusting to 'fast speech' phenomena	Focus on decoding and automatic word/phrase recognition	Focus on grammatical parsing	Integration of linguistic and paralinguistic signals
Word Grab	listening to recognise specific words and phrases	■		■	■	
Shadowing	practising close listening and giving feedback				■	
Race to the Wall	listening for key words and making inferences			■		
Action Skits	listening for details in sequences of action				■	■
Total Recall	listening for specific details in a story	■		■	■	
Bucket List Bingo	listening for specific phrases		■		■	
Map Readers	following a route on a map		■	■		
Details, Details	listening for specific facts (dates, numbers, names, etc.)	■		■		
What's the Line?	hearing 'fast speech', learning phonological rules	■	■			
Pause and Predict	predicting the next word in a story			■	■	■

(Darkened boxes indicate research links to the activities.)

Activity: Word Grab

Introduction

'Word grab' is a highly enjoyable activity that gets students working at the level of word recognition. Because the activity involves physical movement and a game-like format that includes an element of competition, the students tend to be highly engaged while doing it, and engagement is the central concept of active listening. In the second phase of the activity, the students use the words to reconstruct the text. This phase of the activity helps with 'pushed output', an important principle in oral language development.

Aim	• listen for specific words
	• collaborate to 'piece together' a listening passage from isolated words only
	• develop these active listening strategies: selective attention, comprehension monitoring, joint task construction, problem evaluation, summarisation
Level	Beginner +
Time	10 minutes

Materials

Any listening passage of at least a minute's length. Stories are a good genre for this activity.

http://storyteller.net/stories/audio contains audio recordings of fictional stories, fairy tales and myths, told quite slowly and appropriately for language students.

Newspapers and magazines such as *Reader's Digest* often contain human interest stories that are short enough to read to the class.

Preparation

1. Look at the transcript of a listening passage (or listen to the passage noting down key words) and choose around 12–16 important words or phrases. Choose content words such as nouns and adverbs rather than prepositions and articles.

2. Write the words on cards so that they can be grabbed. Prepare to have enough cards to share between three to four students, i.e. if you have twelve people in the class, you will need either three or four sets of cards. If you think you will use the same activity another time and you have the tools, laminate the cards and collect them at the end.

Procedure

1. (before listening) Put the students into groups of three or four. Each group has a set of cards either on the desk or attached to the wall.

2. Tell the class to read the words, and explain that they will hear a passage. When they hear the words on the cards they have to grab them. It is a race.

3. (listening) Play the recording or read the passage.

4. (after listening) At the end, ask the students to count the number of words they grabbed to see who won.

5. Tell the students to work in their groups to try to piece together the story just from the words on cards. This may be very challenging for them.

6. Tell them to listen again for gist.

7. At the end, they try to reconstruct the text by collaborating and putting the words in order.

Variation

- Do the activity after the students have listened at least once so they know the gist and can anticipate which words are coming. This practises a different, but still highly useful, skill.

- Get the students to put the cards in order instead of grabbing them. This is less fun but more collaborative.

- Use some distracters, such as soundalikes, e.g. chores for shores, or homophones, e.g. currants for currents.

Comments

This activity always appeals to kinaesthetic learners – those who learn by moving – and provides a change of pace and style for usually sedentary classes. We tend to emphasise heavily the post-listening reconstruction work, as it acts as a guard against the comment that 'it's just a game'.

Research links

Nation, P. (2008) *Teaching Vocabulary: Strategies and Techniques.* Boston: Heinle ELT.

Schmitt, N. (2000) *Vocabulary in Language Teaching.* Cambridge: Cambridge University Press.

Sample Cards

Advanced

Sample Text

The Ducks that Crossed an Ocean

On 10 January 1992 a container of bath toys was sent by cargo ship from Hong Kong to Washington, USA. While in the middle of the Pacific Ocean, the ship was struck by a storm and the container fell off and broke open. The result was that 29,000 bath toys began a journey across the world's oceans. The toys, which include yellow rubber ducks, green frogs and blue turtles, are called Floatees, and for good reason: they are designed to float. This allowed them to survive the ocean's currents. About 10 months after the original spill, the toys made their first landfall, washing up on the shores of Alaska. Then they began turning up all over the Pacific: Australia, Japan, Hawaii and North America.

Besides spawning two children's books about their amazing journey, it turns out that the toys have other uses. An oceanographer called Curtis Ebbesmeyer realised that Floatees could be useful tools for tracing ocean currents, and so he began tracking them. He predicted correctly that some of the toys would arrive on the shores of Washington State, USA in 1996, and suggested that some would be trapped in the Arctic ice pack, where they would drift eastward at a rate of one mile per day.

In 2003, the makers of the toys – a company called The First Years – offered a $100 savings bond to anyone who recovered a Floatee in New England, Canada or Iceland. Recovery of the toys soared in 2004.

As the story of the Floatees' epic journey became widely known, the toys became collectors' items. They have been sold for up to $1,000 at auctions, and are subject to forensic tests to check they are the real thing. Over 20 years later, they have not all been recovered – many are still floating across the world's oceans. As for the condition of the Floatees, apparently the ducks have lost their yellow colour due to the bleaching of the sun and seawater, but the frogs and turtles remain heroically green and blue.

Activity: Shadowing

Introduction

An active listener listens closely to the speaker, monitoring carefully what is said. One way to encourage close listening is the practice of **shadowing**, repeating all or part of what the speaker has said. This activity presents a framework for practising shadowing in a communicative way, so that the shadowing contributes to the conversation. This activity helps students focus on grammatical parsing, and helps expand their short-term memory in the L2.

Aim
- promote close listening
- practise appropriate forms of feedback
- develop these active listening strategies: repetition, backchannelling, selective attention, seeking confirmation

Level   Beginner +

Time   20 minutes

Materials

None

Preparation

Select a few things, people or places that represent you: perhaps a childhood hero, a person you admire, a place you love visiting, an item of food you consider a luxury. You will use these allusions when you model the activity.

Procedure

1. (before listening) Tell the learners they will be having dialogues in pairs. The topic is: talk about something that represents you – a person, an object, a sound, a taste, an animal, a place. Give your own personal example, if you wish. Give the learners a couple of minutes to prepare.

2. Explain that the listener is to shadow what the speaker says, either repeating the speaker's words, or paraphrasing (saying the same meaning in your own words).

3. Model the activity with a student. For example:

   A (teacher): I chose a lion. A lion represents me.
   B (student): A lion represents you.
   A: Yes, right. I think a lion is strong.
   B: A lion is strong.
   A: Yes, and, for this reason, all of the other animals respect the lion.
   B: Other animals have respect for the lion.

4. (listening) Have the students work in pairs. Give each speaker one minute, and then announce, 'Change speakers'. After each student has had a turn speaking and listening, rotate to new partners. Continue with three new partners.

5. (after listening) Ask the students what they thought of the activity. Did the shadowing help them listen better? Did the shadowing help the speaker? Did it bother the speaker? How can you adjust the shadowing to make it more useful?

6. Pass out the worksheets. Demonstrate the six types of shadowing: full, key word, last word, paraphrase, confirmation and silent. Ask the students to work in pairs to identify which kind of shadowing the listener is using. Answers: (1) key word; (2) full; (3) last word; (4) silent; (5) confirmation; (6) paraphrasing.

7. Go over the answers with the class. Ask the students to reflect: which kinds of shadowing would you like to try?

Variation

Key word shadowing. There are multiple forms of shadowing you can use to encourage close listening, including 'silent shadowing' in which the listener mentally repeats what the speaker says, but not aloud. Another is 'key word shadowing' in which the listener repeats only the key word or phrase from the speaker's turn.

Comments

We have found that shadowing is a useful technique for students who are very reticent to speak in more open discussion activities. By practising shadowing in authentic communicative settings, in which the students are presenting their own ideas and opinions, students gain confidence. Shadowing, as a form of 'elicited imitation', is also used on some measures of spoken language proficiency.

Research links

Fujimoto, D. (2009) 'Listener responses in interaction: a case for abandoning the term, backchannel', *Osaka Jogakuin University Research Repository*, 37, pp. 35–54. http://ir-lib.wilmina.ac.jp/dspace/bitstream/10775/48/1/03.pdf.

LoCastro, V. (1987) 'Aizuchi: a Japanese conversational routine' in Smith, L. (ed.) *Discourse Across Cultures*. New York: Prentice Hall.

Murphey, T. (2000) *Shadowing and Summarizing*, NFLRC Video #11. Honolulu: University of Hawaii, Second Language Teaching and Curriculum Center.

Reidsma, D. *et al.* (2010) *Continuous Interaction with a Virtual Human*, pp. 24–38. Amsterdam: ENTERFACE 2010. http://eprints.eemcs.utwente.nl/16111/01/e10ci_endreport.pdf.

Worksheet

Exercise: What type of shadowing is the listener (B) using?

- Full shadowing (repeating everything you hear, almost like a parrot).
- Key word shadowing (repeating the speaker's key words or phrases).
- Last word shadowing (repeating the final word, phrase, or thought of the speaker, including a very close paraphrase).
- Paraphrase shadowing (repeating an idea, using your own words, but trying to stay close to the speaker's original meaning).
- Confirmation shadowing (asking a confirmation question about what the speaker just said).
- Silent shadowing (mentally repeating what the speaker says, but not aloud).

(1) A: There are four people in my family.
   B: Four people? _____ **shadowing**

(2) A: My mum, dad, my older brother, and me.
   B: Mum, dad, brother, you. _____ **shadowing**

(3) A: This is my older brother, Jin.
   B: Your brother, Jin. _____ **shadowing**

(4) A: He loves dancing.
   B: Mm-hmm. _____ **shadowing**

(5) A: And he talks a lot. He loves talking with his friends.
   B: So you mean he's outgoing.? _____ **shadowing**

(6) A: Yes. And he's never at home. He's always out somewhere with his friends.
   B: He's always going places with his friends. _____ **shadowing**

Activity: Race to the Wall

Introduction

Many students can be classified as kinaesthetic learners. They enjoy learning through movement because they find physical activity stimulating and they are more engaged when asked to move. This activity challenges the listener to focus on small details of the passage, such as key words and names, in order to find out where to go in the classroom. Apart from its entertainment value it also activates students' general knowledge and asks them to exercise their memory and oral summarising skills during the follow-up feedback stage.

Aim
- listen for key points
- use physical action as a sign of understanding
- develop post-listening skills of paraphrasing and checking information
- develop these active listening strategies: predictive inferencing, repetition, selective listening

Level   Beginner +

Time   15 minutes

Materials

Five large sheets of paper, each labelled as a continent.

To find information about countries, go to the country profiles on the BBC website: http://news.bbc.co.uk/2/hi/country_profiles/default.stm.

The topic need not necessarily be countries. See variations below:

- Animals – the categories could be domestic animals, ocean-dwellers, wild cats, marsupials or insects.
- Food – fruit, vegetables, meat, desserts, main courses.
- Games – board games, single-player games, team games, card games, dice games.

Preparation

1. Select the items of vocabulary (in this case, countries) to review.

2. Be ready to talk about the items, including facts and opinions.

Procedure

1. (before listening) Label the four walls and the floor of the classroom using large sheets of paper. Each wall, plus the floor, is thereby given the name of one continent.

2. (Optional) Elicit from the group as many countries as they can name from each continent and one or two pieces of information about each country, e.g. what it's famous for.

3. If necessary, change the layout of the furniture so the students have a clear physical pathway to each of the walls.

4. Tell the students to stand in the middle of the room.

5. Explain that they will hear some information about a country. When they know which country it is, they should go to the continent to which it belongs, and attach themselves with three parts of their body (e.g., two hands and foot) to the correct wall/floor.

6. (listening) Demonstrate the activity by doing one example (you can use one of the sample scripts). Stress that the students must move immediately when they know which country is being described.

7. (Optional) Give points for the winner or some kind of token, such as a sticker.

8. After descriptions of each country, elicit the answer from the whole class, then have the students pair up for 30 seconds, saying what they remember from the teacher's description of the country.

9. (after listening) Ask the students what they remember. Discuss briefly. Then get the students to return to the middle of the room.

10. Begin the next monologue about a different country.

Variation

- For a purely bottom up exercise that uses the same principles (the version above contains the top down element of students using general knowledge to guess the answers), the teacher simply says the target words without embedding them in a stream of speech. This practises word recognition.

- The teacher numbers the walls (e.g. Wall 2, Wall 3, etc.). Then the teacher says target phrases (e.g. where are you going?; a bucket of ice; gin and tonic, etc.). If the phrase has three words, students run to Wall 3; if the phrase has four words, students run to wall 4, etc. This variation asks students to recognise word boundaries, segmenting the stream of speech into individual words, and practises coping with features of connected speech such as elision and assimilation.

Comments

We use this activity with all kinds of classes – adult General English, Business English, ESP, Young Learners – as the content is easy to manipulate and grade, according to our students' level and interests. Also the activity takes away part of the stress of listening because it is enjoyable. This helps to lower the **affective filter**.

Research links

Gardner, H. (1983) *Frames of Mind: The Theory of Multiple Intelligences*. New York: Basic Books.

Sample Script

Audio File 9: Race to the Wall

Please visit www.routledge.com/9781408296851 for an audio recording of this transcript.

Countries

> **Advanced level:** This country had an extensive empire that began to decline in the twentieth century. At the height of its power, it presided over huge areas of the world, including India and large parts of Africa. It's a sports-loving nation that has won both the football and the rugby world cups once, and is one of the best at cricket. One of its greatest exports is its music ranging from The Beatles through the Rolling Stones to Coldplay.

(Answer: the United Kingdom)

Food

> **Intermediate level:** This is a type of dish that originated in Greece, though most people associate it with Italy. It's made with a flat round piece of bread topped with tomato sauce, cheese and various vegetables and types of meat or seafood. Then it's baked in the oven.

(Answer: pizza)

Animals

> **Beginner level:** It's an animal with four legs and a tail. It sometimes lives in the home. It is good at smelling things, and the police use it.

(Answer: dog)

Activity: Action Skits

Introduction

Total Physical Response (TPR) is a 'listening first' or 'natural approach' methodology that keeps students active, but does not demand much speaking. TPR involves students listening to the teacher and performing actions, thereby acquiring the basics of the language without the necessity of speaking. When this method works, it tends to promote automatic word recognition. The students begin to think in the target language and respond naturally to language in the L2 without the need for translation. This activity utilises the principles of TPR for understanding oral commands, and combines it with controlled speaking and drama techniques.

Aim
- become comfortable listening without needing to respond verbally
- focus on physical details as you listen
- use your body movements to help you understand
- develop these active listening strategies: contextual inference, predictive inferencing, emotional monitoring, seeking clarification, performance evaluation

Level Beginner +

Time 10 minutes

Materials

None needed, but you may wish to warm students to the activity by showing some examples of silent skits, for example:

In a cinema: http://www.youtube.com/watch?v=GsVcC0za1dA

Or you may wish to give some 'miming' lessons to the group to increase awareness of how to communicate through movement, for example:

Mime Basics: Acting Tips and Techniques. How to Walk in Place in Mime: http://youtube.com/watch?v=jTe0q4UFvn0.

Preparation

Make a list of verbs that will serve as instructions for the 'actors'. (They will perform a skit.) Break down the actions into as many specific verbs as possible. Include verbal actions such as 'ask', 'answer', 'say', 'tell', 'react', 'respond' (see the worksheet samples).

http://www.ehow.com/video_2261229_create-story-mime.html

http://www.amazon.com/Mime-Acting-Production-Principles-ebook/dp/B004YDMDWM

If students have an interest in creating dramatic skits, here are some links to explore:

http://www.bencrockerpantomimes.com/s

This site includes scripts like *Jack and the Beanstalk* and *Sleeping Beauty*.

Procedure

1. (before listening) Do a short physical warm up – standing, walking, stretching – to get students on their feet and moving about.

2. Introduce the topic and the setting for your action skit. For example, 'You are a parent. You have a one-month-old baby. Ready?'

3. (listening) Read the sequence of actions in a natural rhythm, but leave long pauses for students to understand and perform the actions. If most students do not understand a step, repeat it and, if necessary, mime the action yourself.

4. Continue through to the end of the skit.

5. (after listening) Have the students work in pairs to create a mime scene, involving 10 or so steps, and 1 or 2 characters. Encourage the students to think in terms of a beginning (of 3 or 4 steps), a middle, and an end, each with 3 or 4 actions. Each step needs to use an imperative verb.

Variation

Team mime. Put the students into two teams. The teams line up facing each other and number off. The idea is to give an instruction to the person opposite, who must listen and follow the instruction either by miming or doing it. Student 1 in Team A gives an instruction to Student 1 in Team B. Student 1 in Team B then gives an instruction to Student 2 in Team A. Student 2 in Team A then gives instructions to Student 2 in Team B, etc. Continue until everyone has had at least two turns of giving and receiving instructions.

	Action	Mime	Mime sequence
Beginner	Stand up. Sit down. Touch your head. Point to the door.	Put on a hat. Play the piano. Make a phone call. Cut the bread.	1. Wake up. 2. Get dressed. 3. Make coffee. 4. Brush your teeth.
Intermediate	Shake hands with a partner. Scratch your forehead. Turn around, jump once, clap twice. Wave your arms for attention.	With a partner, give and receive a present. Play an instrument in an orchestra. Do a bungee jump. Play a board game.	1. Put up a tent. 2. Build a camp fire. 3. Plan a route on a map. 4. Reach the top of a mountain.
Advanced	Walk with a limp. Greet a partner like a long-lost friend. Stretch a muscle that feels stiff. Act like a spy.	Teach your partner to operate a machine. Build a shed. Take out someone's tooth. Mime something beginning with . . . a, b, c, etc.	1. Dig into the ground and discover something. 2. Clean and examine it. 3. Examine it in the lab using technology. 4. Make a speech about it and answer questions.

Comments

We have found that using listening activities that require minimal verbal response can help learners relax and focus on listening development. This type of TPR activity can be used from time to time as a warm-up exercise, or it can be incorporated into other activities.

Research links

Asher, J.J. (1966) 'The learning strategy of the Total Physical Response: a review', *The Modern Language Journal*, 50(2), pp. 79–84.

Asher, J.J. (1969) 'The Total Physical Response approach to second language learning', *The Modern Language Journal*, 53, pp. 3–17.

Asher, J.J. (2003) *Learning Another Language Through Actions* (6th edition). Los Gatos, CA: Sky Oaks Production.

One Stop English: http://www.onestopenglish.com/support/methodology/teaching-approaches/teaching-approaches-total-physical-response/146503.article

TPR World: http://www.tpr-world.com/

Blaine Ray Total Physical Response Stories: http://www.blaineraytprs.com/.

Worksheet

Audio File 10: Action Skits

Please visit www.routledge.com/9781408296851 for an audio recording of this transcript.

Skit for one person: Baby

1. *Open* the door quietly.
2. *Flip on* the light switch.
3. *Tip toe* over to the baby's crib.
4. *Pull back* the blanket from your baby.
5. Carefully *pick up* your baby.
6. *Cradle* the baby in your arms.
7. *Rock* the baby gently.
8. *Hum* a lullaby to the baby. ('Rock-a-by baby . . .')
9. *Pull* the baby close to your neck.
10. *Pat* the baby's bottom.
11. Uh-oh. You *smell* something.
12. Uh-oh. Baby's wet. *Pull* the baby away from your body. Go to change the baby's nappy.

Skit for two or three people: Hotel

1. This is a hotel. (Name), you're the clerk and (Name), you're a tourist. Clerk, *stand* behind the counter. Tourist, *enter* the hotel revolving door. *Pull* your suitcase behind you.
2. Tourist, *walk* up to the counter and *smile* at the clerk.
3. Clerk, *smile* at the tourist and *say*, 'Yes, may I help you?'
4. Tourist, *pull* your suitcase beside you, and *say*, 'Yes, I have a reservation.'

5. Clerk, *ask*, 'Can I have your name, please?'

6. Tourist, *tell* the clerk your name.

7. Desk clerk, *look* down at your computer screen and *punch* a few buttons. *Frown* and *say*, 'I'm sorry. We don't have anybody by that name.'

8. Tourist, you're a little worried. *Lean* forward and *try to see* the computer screen. *Say*, 'Are you sure?'

9. Clerk, *scan* the reservation list on your screen again and *ask*, 'How do you spell your last name?'

10. Tourist, *pronounce* your last name again and then *spell* it slowly.

11. Clerk, *clear* your throat and *point* at the name on your computer screen and *say*, 'Oh, here it is. I'm sorry.'

12. Tourist, you feel relieved. *Sigh* and *say*, 'Oh, good.'

13. Desk clerk, *request* the tourist's passport. *Ask*, 'Would you show me your passport please?'

14. Tourist, *take* your passport out of your pocket and *hand* it to the clerk.

15. Clerk, *compare* the name on the card with the name on the passport to make sure they're the same. Then *give* the passport back to the tourist.

16. *Take* the passport and *place* it back in your pocket.

17. Clerk, now *place* the tourist's room key into a folder, *hand* it to the tourist and *say*, 'Here you are, Mr/Ms _____. Sorry for the trouble. I've upgraded you to a luxury suite.'

18. Tourist, *nod*. *Take* the key and *thank* the clerk.

19. Clerk, *motion* to the bellhop with your hand. *Point* to the tourist's bag and *say*, 'Please show Mr/Ms _____ to his/her room. He/She will be staying in the Presidential Suite.'

20. Bellhop, *take* the bag and *lead* the tourist to the elevator.

Activity: Total Recall

Introduction

Recall is one aspect of listening that is often given the most weighting in formal assessment. Developing recall is, of course, an important part of listening development, so most learners tend to value instruction that challenges their memory. This activity presents a framework for assessing recall that will give students practice with standardised question types and also with rating their confidence level. This type of monitoring is a useful strategy that can help learners focus on details as they listen.

Aim	• improve attention to detail
	• practise different forms of assessment
	• develop these active listening strategies: question elaboration, repetition, performance evaluation, comprehension monitoring, double-check monitoring
Level	Beginner +
Time	15 minutes

Materials

Choose a short (about three minutes) contextualised audio or video segment. The worksheet uses *Bus Love*, a short video drama made for ESL learners: http://lingual.net/2011/02/28/bus-love/

Other possible sources include:

- Narrative: clips from Hollywood movies, hosted at *Learn English Feel Good*; this clip from *Hereafter*: http://www.learnenglishfeelgood.com/eslvideo/esl_movieclip2.html

    http://www.youtube.com/watch?v=WvqDqi8-Mpc&feature=player_embedded

- Expository: sample lectures from simulated TOEFL tests at englishclub.com http://www.englishclub.com/esl-exams/ets-toefl-practice-listening.htm

- Short conversations: Small talk topics from EslFast.com http://www.eslfast.com/robot/topics/smalltalk/smalltalk.htm

Preparation

Preview the film or audio clip you will be using. Obtain (or transcribe) a complete script, if possible. Create an exercise that tests recall of the clip, including some very specific detail questions. You can test recall of the story line, situational details, of language spoken, non-verbal communication signals, or 'action between the lines' (the intent of the actors).

Procedure

1. (before listening) Preview the topic of the clip you will be using. Introduce some of the key vocabulary and give a short summary of the storyline (but don't give away the plot!).

2. Tell the students they will be given a recall test at the end. They *can't* take notes; they are just to watch and listen carefully.

3. (listening) Play the clip. If necessary, you can pause the clip to allow learners to reflect on what they have seen up to that point.

4. (after listening) At the end of the first viewing, pass out the exercise sheets (see the worksheet). Students can work alone or in groups.

5. Have students complete their answers. For each answer, they should also rate their 'confidence'. How sure are you? (0 = no idea, complete guess; 1 = a slightly educated guess, feeling lucky; 2 = a pretty sure guess, my gut tells me this is right; 3 = an absolutely certain answer, I'd bet my life on it)

6. After students complete the exercise on their own, have them work in pairs to compare answers and confidence ratings.

7. Play the extract one more time to confirm answers.

Variation

Silent comprehension. To take listening out of the equation – and to show that a good deal of comprehension and recall is visual and 'schematic' (what we already know from experience with similar situations) – try a recall exercise with the sound turned off. Focus solely on visual aspects of comprehension. Almost any video clip will work for this. If you'd like to try a video clip that was used for cross-cultural research, try the non-verbal film *The Pear Story* (available at http://pearstories.org/).

Comments

We have discovered that including a self-rating of 'confidence' is a useful technique for developing a monitoring strategy that is associated with advanced level listening. Monitoring enables you to focus – to direct your energy – and to find out where you need to put in additional effort to clarify what you don't understand, or where you need to compensate or simply guess.

Research links

Lynch, T. (2006) 'Academic listening: marrying top and bottom' in Uso-Juan, E. and Martinez-Flor, A. (eds) *Current Trends in the Development and Teaching of the Four Language Skills*. Berlin: DeGruyter.

Wilson, M. (2003) 'Discovery Listening – improving perceptual processes', *ELT Journal*, 35, pp. 325–33.

Worksheet

Read each statement. Is it True (T), False (F), or No Information given in the film (N)? Are you: absolutely confident (3); pretty sure (2); umm, maybe (1); it's a complete guess (0).

	T / F/ N	How confident are you?
1. The title of the film is *Bus Stop*.		
2. Pete is wearing a brown T-shirt.		
3. The number on the Bus is 14.		
4. Pete inserts a five-dollar bill into the slot.		
5. The bus driver is wearing glasses.		
6. Penny is sitting next to a young woman.		
7. Penny is wearing a purple vest.		
8. Pete puts his transfer into his trouser pocket.		
9. Penny has blonde hair.		
10. Pete sits in the seat in front of Penny.		
11. Penny talks to Pete first.		
12. Pete tells Penny he's a student.		
13. Penny says, 'I got food poisoning there one time.'		
14. The seats on the bus are red.		
15. Penny says that she works with kids.		
16. Penny asks Pete, 'Where do you live?'		
17. Pete misses his stop.		
18. Penny writes her phone number on her business card.		
19. When Pete gets off the bus, he feels depressed.		
20. Pete loses Penny's phone number.		

Activity: Bucket List Bingo

Introduction

This activity uses an idea from a film called *The Bucket List* and a popular group game as the basis for an enjoyable personalised activity. Bingo requires its players to listen only for certain items (numbers or, in this case, words) – a selective listening process – in order to win. This activity may seem to be useful only for beginners but, depending on the items on the bingo card, and the complexity of the input, the game can be useful for learners of all proficiency levels.

Aim	• listen for specific phrases
	• talk about your wish-list
	• develop these active listening strategies: advance organising, selective attention, noticing attention, comprehension monitoring
Level	Beginner +
Time	30 minutes

Materials

None, but see synopsis of movie *The Bucket List*, for more context related to the activity:
http://en.wikipedia.org/wiki/The_Bucket_List

Preparation

Prepare to talk about your 'bucket list', a wish-list of things you'd like to do in life.

Procedure

1. (before listening) Draw a grid on the board with nine verbs, one per square, that represent your 'bucket list' (see sample visual). This is a list of things you'd love to do in life, e.g. visit Tibet, eat fresh lobster, go to Carnival in Rio de Janeiro, learn to dance the tango, etc. Ask the students to Copy the grid and to circle any three of the verbs.

2. (listening) Explain that you'll now talk about the actions and why you'd like to do them. The students' task is to play bingo: they draw a big X through their circled sections as soon as they hear you describing them. The first person to 'X' all three of their circled actions shouts 'bingo!' and is the winner.

3. Even if there is already a winner, continue the descriptions of your bucket list until you have finished.

4. (after listening) Put the students into small groups and ask them to complete the verb phrases in the grid (e.g. learn to . . . *play the violin*), and to talk through what they can remember about the teacher's bucket list.

5. As a follow-up, get the students to write their own bucket list of five to seven items. You may wish to point out the verbs you used (eat, go, visit, play, etc.) as a guide to the type of things for them to include.

6. In groups, the students talk about their bucket list.

Variation

- Instead of playing 'bingo', the students tick the things they'd also like to do, and number the items in the order you describe them. Then they follow stages 4–6.

- Include the full verb phrases in the grid (*learn to play the violin, climb up Mount Kilimanjaro*, etc.). This makes for an easier task for the students.

- To make this a top down instead of bottom up activity, the students look at the verbs in the grid and guess what the speaker will say to complete the phrase, i.e. on seeing the verb *meet*, the student might say, 'I think you'd like to meet a famous soccer player: Lionel Messi!'

Comments

This activity involves extended listening and speaking, but we have used bingo for many other activities, e.g. focusing on pronunciation by having different phonemes in the grid, with the students ticking the phonemes they hear.

Research links

http://www.teacherjoe.us/TeachersSpeakingBingo.html.

http://www.eslactivities.com/bingo.php.

Sample Visual

learn to	*climb up*	*watch a*
eat	*meet*	*go*
do	*speak*	*sing*

Activity: Map Readers

Introduction

Map readers is a very flexible activity. The teacher-generated input can include any type of place on a map: from countries to local attractions, depending on the map. The use of real-life places and real maps tends to be very motivating for students as they are using material and skills that transfer immediately to the world outside the classroom. The activity is framed as a bottom up activity in that the task revolves around pointing out places on a map. Alternatively, teachers can use a more extensive task such as asking the students to listen for information about the places.

Aim   • listen for names of places

   • develop these active listening strategies: selective listening, world elaboration, visual elaboration, seeking clarification

Level   Intermediate +

Time   15 minutes

Materials

One map per student, that includes the places you will talk about.

Possible sources:

http://www.mapsofworld.com/
http://geography.about.com/library/maps/blindex.htm
http://maps.google.com/

Preparation

Prepare to talk about several places of interest. Your talk needs to include at least five places, and should be graded for your students to be able to follow.

Procedure

1. (before listening) Hand the students a map that includes the names of places you are going to talk about.

2. Tell them their listening task is to mark the places you mention on the map.

3. (listening) Begin a monologue about the places you have visited. Include some repetition and rephrasing in order to allow the students multiple opportunities to find the places on the map.

4. (after listening) Check their answers and ask for further information about what you said.

Variation

- **Follow the journey**. Change the input and have students trace a journey on a map, drawing lines to signify movement. For example, the teacher could describe a trip around the Caribbean and the students, following the journey, draw lines between different countries or islands, e.g. from Cuba to Haiti.

- **Two-speed listening**. Try first delivering the input (your descriptions of places) at a slightly slower speed than normal. Then repeat the input at a rapid, natural speed. This trains the students to listen to fast speech. Some teachers have used Audacity – a free digital audio editor – to slow down sections of recordings before playing them at full speed afterwards.

Comments

When working in ESL situations, where the teacher is in their home town and the students are foreigners, we have often done this activity describing local attractions, and using a local map. This introduces new students to their local environment and gives them ideas for places to visit.

Research links

http://busyteacher.org/2842-realia-esl-classroom.html.

http://www.teachingenglish.org.uk/language-assistant/teaching-tips/realia.

Activity: Details, Details

Introduction

In this activity, the students listen for very specific information and then begin to 'see the big picture' through the gradual build-up of this data. While listening for names, times, dates and numbers, the listener will be making connections between the details. This type of mental engagement is desirable, as our brains always try to forge connections between such data. So even though the listening task demands focused attention on smaller units of information, the students gradually will be forming a hypothesis about the overall meaning of the input. This is a type of interaction between top down and bottom up processing.

Aim
- focus on discrete units of information
- collaborate to build a whole story based on small units of information
- develop these active listening strategies: selective listening, repetition, noting, retrospective inferencing

Level Intermediate +

Time 15 minutes

Materials

A factual recording, e.g. the history of something, a radio programme, or a story. The recording needs to have names, times, numbers and dates in sufficient quantity and of sufficient importance to the story.

Possible source for higher level groups:

http://www.npr.org/ (e.g. morning edition of the news)

http://www.bbc.co.uk/news/video_and_audio/ (BBC's one-minute world news summary)

http://www.cnn.com/video/index.html (CNN's collection of short news clips)

or see the sample text for an alternative.

Preparation

Listen to the recording and note down times, names, dates and numbers contained in it.

Procedure

1. (before listening) Tell the students they are listening for names, times, numbers and dates only. As they listen, they must write these down.

2. (listening) After a first listening, the students compare answers.

3. They listen again to refine stages 1 and 2.

4. Include a clarification stage, giving the students the opportunity to ask you questions: 'Did you say . . . ?' 'How many were there?' etc.

5. (after listening) In plenary, the class says the names, times, dates and numbers, which the teacher writes on the board. From this they try to piece together the story. The teacher guides them as they speculate: 'OK, so President Obama went where? And for how many days?'

6. The students listen a third time for confirmation.

Variation

- For high-level groups, use a radio news programme. The headlines, as well as the main stories, tend to be full of names, dates and numbers.

- The students note down numbers, dates, etc. as the teacher gives a list of facts about countries of the world. A good source is http://www. worldatlas.com/geoquiz/thelist.htm The students could be given a map of the world as visual support.

Comments

Listeners normally listen for gist first. Asking them to listen for names, times, dates and numbers first is challenging for students because it subverts their normal listening process. We have found, however, that students tend to enjoy both experimenting with their listening processes as well as piecing together the story.

Research links

Buck, G. (2001) *Assessing Listening.* Cambridge: Cambridge University Press.

Sample Text and Script

Continents (by size)

1. Asia (44,579,000 sq km)
2. Africa (30,065,000 sq km)
3. North America (24,256,000 sq km)
4. South America (17,819,000 sq km)

5. Antarctica (13,209,000 sq km)
6. Europe (9,938,000 sq km)
7. Oceania (7,687,000 sq km)

Continents (by population; 2006 est.)

1. Asia (3,879,000,000)
2. Africa (877,500,000)
3. Europe (727,000,000)
4. North America (501,500,000)

5. South America (379,500,000)
6. Australia/Oceania (32,000,000)
7. Antarctica (0)

Continents (by the number of countries)

1. Africa (54)
2. Europe (46)
3. Asia (44)

4. North America (23)
5. Oceania (14)
6. South America (12)

Country population (largest; February, 2006 numbers)

China 1,306,313,800
India 1,080,264,400
USA 295,734,100
Indonesia 241,973,900
Brazil 186,112,800

Pakistan 162,419,900
Bangladesh 144,319,600
Russia 143,420,300
Nigeria 128,772,000
Japan 127,417,200

Country population (smallest; February, 2006 numbers)

Vatican City 920
Tuvalu 11,640
Nauru 13,050
Palau 20,300
San Marino 28,880

Monaco 32,410
Liechtenstein 33,720
St Kitts 38,960
Marshall Islands 59,070
Antigua and Barbuda 68,720

Oldest countries

San Marino (301 AD)
France (486 AD)   Portugal (1143 AD)
Bulgaria (632 AD)   Andorra (1278 AD)
Denmark (950 AD)   Switzerland (1291 AD)

Youngest countries

South Sudan (July, 2011)   Czech Republic (1993)
Montenegro (July, 2006)   Eritrea (1993)
Serbia (July, 2006)   Slovakia (1993)
East Timor (2002)   Bosnia/Hertzegovina (1992)
Palau (1994)

Richest countries (GNP per person in US dollars)

Luxembourg ($45,360)   Liechtenstein ($40,000)
Switzerland ($44,355)   Norway ($34,515)
Japan ($41,010)

Poorest countries (GNP per person in US dollars)

Mozambique ($80)   Ethiopia ($100)
Somalia ($100)   Congo ($100)
Eritrea ($100)

Source: data taken from www.worldatlas.com

Sample teacher script

OK, the biggest continents? Number one is Asia, which is over forty-four million square kilometres. Second on the list is Africa, which is just over thirty
million square kilometres. What about population? Which are the most
populated continents? Well, the answers are the same. Number one is Asia
and number two is Africa. And, of course, the least populated continent
is Antarctica, with a population of . . . zero! It's too cold to live there. As
for countries, China and India are the most populated, but which is the number three? The answer is the USA, which has around three-hundred million
inhabitants.

Activity: What's the Line?

Introduction

In natural spoken English, a rhythm-timed language, a number of physical transformations take place regarding sounds and words. For example, consonants are assimilated and vowels reduced. This phenomenon of 'fast speech' or 'connected speech' is often confusing for learners, especially those who have learned the citation forms of written words as 'fixed'. It is important for L2 listeners to become aware of how these changes occur, and this activity presents different fast speech patterns, using lines from famous films, in order to help learners develop an awareness of how 'fast speech' sounds are made.

Aim
- notice fast speech patterns
- develop discrimination between citation forms and reduced or assimilated forms
- develop these active listening strategies: noticing attention, comprehension monitoring, repetition, visual elaboration, resourcing (using a phonetic alphabet, diacritics)

Level Intermediate +

Time 15 minutes

Materials

Short clips of natural spoken English. Suggestion: 40 inspirational speeches in 2 minutes (see the worksheet):
http://www.youtube.com/watch?v=d6wRkzCW5ql

Preparation

1. Preview the clips you plan to use. (You don't need to use all of the lines in these large collections. About 10–15 is a good number for one dose of this kind of practice.)

2. Create a worksheet, or set of notes to indicate how many words (blanks) for each quote you intend to use. Make notes of any vocabulary you may wish to pre-teach – or simply provide those words in advance when you show the number of blanks.

3. Reacquaint yourself with common 'fast speech' forms in spoken English.

Form	Example	Process
/d/ + /j/→/dʒ/	would you	assimilation
/d/→Ø	could you	elision
/t/→Ø	can̶t	elision
/t/ + /j/→/tʃ/	get you	assimilation
/h/→Ø	h̶e	elision
/g/→Ø	doin̶g	elision
/k/→Ø	tal̶k	elision
going to → gonna		fused form (co-articulation)
got to → gotta		fused form (co-articulation)
want to → wanna		fused form (co-articulation)

Procedure

1. (before listening) Divide the class into two groups. Keep a score sheet somewhere visible to the whole class.

2. (listening) Play the first clip. These are very short clips. Be careful to pause right at the end of the scene. Write blank lines on the board to indicate the number of words in the clip.

3. Encourage students to call out any words they can identify. Write the words on the correct line. Give a point to the team that called out the word first. (You'll need a scorekeeper.)

4. Play the scene again, until the class has identified the entire phrase or sentence. You can give clues – the first or last letter of a missing word – if needed.

5. Give a bonus point to the team that completes the phrase. Give another bonus point to the team that can identify the movie from which the scene came. (Or, if the movies have been identified in the clip, give bonus points if the students can name the characters.) And, if you like, give an additional bonus point to the team that can paraphrase the line.

6. (after listening) After the class has identified one of the phrases, model the pronunciation as closely as you can to how it was spoken in the clip.

7. Use symbols – underlines and arrows and strikethroughs – to indicate stress, reductions and assimilations.

8. At the end of the game (a preset number of scenes), acknowledge the winning team.

9. (follow up) Have the students work in pairs to practice the lines from the scenes. Note that they should aim to mimic the intonation and attitude of the speaker, *not* the precise pronunciation, with all of the assimilation and reduction. Accurate pronunciation is not needed to improve aural discrimination, which is the goal of this exercise.

Variation

Sound off. Watch the video clip first with the sound off. See if the students are able to recognise the film, work out the relationships between the characters, and perhaps even make up lines for the speakers. Watching with the sound off increases anticipation, and promotes greater top down processing when the sound is on.

Comments

We have found it's easier to work with bottom up processing when you use contextualised materials. Creating interactive games – such as having students work in teams to decipher the clues – also helps fuel motivation to learn more about phonology.

Research links

Brown, J.D. and Kondo-Brown, K. (2006) 'Introducing connected speech' in Brown, J.D. and Kondo-Brown, K. (eds), *Perspectives on Teaching Connected Speech to Second Language Speakers*, pp. 1–15. Honolulu: University of Hawaii Press.

Buck, G. (2001) *Assessing Listening*. Cambridge: Cambridge University Press.

Crawford, M. and Ueyama, Y. (2011) 'Coverage and instruction of reduced forms in EFL textbooks', *The Language Teacher*. http://jalt-publications.org/files/pdf-article/art1_17.pdf.

Worksheet

Script: italicised segments indicate an assimilation or reduction.

1. Will *you* fight?
   No, we *will* run *and* we *will* live.

2. *Shame on* you!

3. This *could be* the *greatest night of our* lives, but you're *going to* let *it be* the worst.

4. And I guarantee a week *won't go* by *in your* life you *won't regret* walking out, *letting them* get the best of you.

5. Well, I'm not *going home.*

6. *We've* come too far.

7. And I'm *going to* stay right here and *fight for* this lost cause.

8. A day may come when the courage *of men* fails, but it is not *this day.*

9. The line *must be drawn here.* This far, no further.

10. I'm not *saying* it's *going to be* easy.

11. You're *going to work* harder *than you've ever worked before.*

12. Well, that's fine. *We'll just get* tougher with it.

13. If a person *grits his* teeth *and shows* real determination.

14. Failure is *not an option.*

15. *That's how* winning is done.

Activity: Pause and Predict

Introduction

As its name suggests, this activity explicitly practises the listening strategy of predicting. As we hear a string of words, we develop an ability to antici-pate the next words based on our knowledge of the language (grammar, vocabulary and speaking styles) as well as our general knowledge of the world. Anticipating – or activating possible words and phrases in short-term memory – is something that all successful listeners do, so it is important to demonstrate this strategy to students overtly.

Aim
- listen intensively
- develop awareness of collocations, chunks and lexical-grammatical patterns
- develop these active listening strategies: predictive inferencing, linguistic inferencing

Level Beginner +

Time 5–15 minutes

Materials

Any piece of spoken discourse that contains numerous examples of com-mon collocations and chunks.

Preparation

1. Find an interesting passage to read to the students or a good, clear recording with a transcript.
2. Mark at least 12–16 places on the transcript where your students should be able to make a guess at anticipating the next word or phrase.

Procedure

1. (before listening) Preview the topic of the listening text and set one or two gist questions.
2. (listening 1) Read or play the text and have the students work together to summarise orally the main idea of the passage.

3. (listening 2) Tell the students they will hear the passage again, but this time with pauses. When there is a pause, they need to shout out the word that comes next. Do an example.

4. Read the passage at normal speed, pausing at the parts you marked. The students shout out the next word. Give credit if they shout out a word that is possible, even if it is not the correct word in this particular text.

5. (after listening) Congratulate the class on its ability to anticipate words. The skill is not only about remembering the passage but also about developing language awareness.

Variation

The teacher reads aloud with a student in the other role. They pause at the end of their turn, and students predict what the conversation partner will say.

Comments

We use this activity after all kinds of listening passages, and we have found that it works for all types of class. The students tend to enjoy it because they have a chance to demonstrate their ability in a semi-competitive way in front of the class. This motivates them to focus intensely in order to 'get' the answers.

Research link

Field, J. (1998) 'Skills and strategies: towards a new methodology for listening', *ELT Journal*, 52, pp. 110–118.

Sample Script

Audio File 11: Pause and Predict

Please visit www.routledge.com/9781408296851 for an audio recording of this transcript.

Intermediate/Advanced

Nollywood

First there was Hollywood, then Bollywood in Bombay, India, and now there is Nollywood, one of Nigeria's booming industries. The Nigerian film industry is the world's third largest. The films themselves are not notable for their quality. They are low-budget films, with even lower production values, but they are hugely popular in Africa.

The actors often learn their lines just a few minutes before filming starts. That is *if* filming starts. Power cuts are common in Nigeria and they cause delays of hours. Then there are the traffic jams, meaning actors and crew often arrive late. Then there is the noise. Street vendors are constantly walking by, shouting to the public to buy their goods. Film crews either have to pay them to walk a different route or hire armed police.

Around 2,000 films are made every year, mostly in the capital Lagos, and not for the cinema; instead, they go straight to video. The films are made cheaply and quickly. Most cost only 15–20 thousand dollars, and are finished in a couple of weeks. Then the censors get to work on them; the bad guys are never allowed to escape. If a criminal has a happy ending, it has to be re-filmed, that is if the actors haven't already started work on another movie.

Nollywood: marked-up version

First there was Hollywood, then Bollywood in Bombay, India, and now there is Nollywood, one of Nigeria's booming industries. The Nigerian film industry is the world's third / largest. The films themselves are not notable / for their quality. They are low-budget films, with even lower production / values, but they are hugely / popular in Africa.

The actors often learn their / lines just a few minutes before filming starts. That is *if* filming starts. Power cuts are common in Nigeria and they cause delays of / hours. Then there are the traffic / jams, meaning actors and crew often arrive late. Then there is the noise. Street vendors are constantly walking by, shouting to the public to buy their / goods. Film crews either have to pay them to walk a different route or hire armed / police.

Around 2,000 films are made every / year, mostly in the capital Lagos, and not for the cinema; instead, they go straight to / video. The films are made cheaply and quickly. Most cost only 15–20 thousand dollars, and are finished in a couple / of weeks. Then the censors get to work on them; the bad guys are never allowed / to escape. If a criminal has a happy ending, it has to be re-filmed, that is if the actors haven't already started work on another / movie.

Frame 4
Interactive Frame

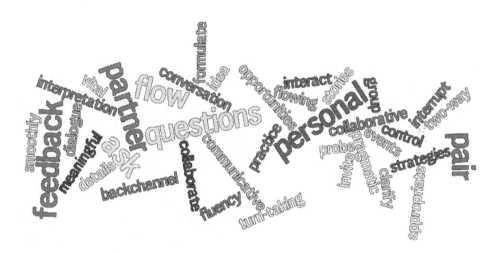

Introduction

The **Interactive Frame** deals with listening in conversations and groups. Interactive listening is concerned with collaborating with others, actively trying to understand what they say, clarifying the ideas they are formulating, and giving meaningful feedback. In effect, interactive listening means holding up your side of the conversation and actively co-constructing a dialogue with the speaker, to the extent this is possible and appropriate.

This frame addresses questions such as: How can I get my students to interact more? Can I help them develop confidence and techniques to become more active as listeners in conversations? How can I get students to become more curious about what other people say? Can I help my students identify what parts of the interaction are giving them difficulties? What can I do to get students to 'repair' interactions that are not going well? Is there anything I can do to raise awareness about the influence of the students' interaction style in their L1, so that they can choose to alter their interaction style in their L2?

Research shows that pragmatic competence – the ability to interact and to participate actively in conversations – is valued as a key component of L2 proficiency. This frame is concerned with developing that aspect of language proficiency. With all activities in this frame, it is helpful to have learners focus on developing strategies of planning, collaborating and monitoring. (See Appendix 1 for a list of active listening strategies and examples of specific strategies the students can try out.)

Ten illustrative activities

The ten activities presented in the Interactive Frame are:

Activity	Description	Creating new learning opportunities	Improving quality of interaction	Focus on listener response	Focus on pushed output	Focus on learner-initiated interaction
Photoshop	finding differences in a photo-shopped picture		■	■	■	
Whisper Dictation	listening carefully in difficult circumstances			■		
Interrupted Story	interacting with a speaker to get a story straight	■				■
Interactive Quiz	interacting through teacher–student questions		■		■	
Blind Forgery	drawing based on descriptions of artworks	■				■
Milestones	group sharing of autobiographical material			■		■
My Turn/ Your Turn	reconstructing an extract in pairs		■		■	
Guest Speaker	interacting with guest lecturers			■		
Paraphrase	paraphrasing as a form of feedback			■	■	
Pecha-Kucha	presenting in front of an active audience	■				

(Darkened boxes indicate research links to the activities.)

Activity: Photoshop

Introduction

Like most activities in this frame, *Photoshop* involves interactive listening and speaking. The task goal of identifying differences between two pictures can be achieved only through collaboration and attentive listening. Interaction is the key. Through interactive activities such as this, learners gain valuable practice in creating symmetry in conversations. In other words, a type of dance takes place between speaker and listener as they swap roles and take on the burden of initiating the next move. To make this work, the listener must be able to control the pace of the input, keep focusing on the task goal, and help the partner through using clarification expressions, repetition and rephrasing.

Aim • interact with a partner to discuss a picture

• listen for detail

• develop these active listening strategies: selective listening, backchannelling, seeking clarification, joint task construction, world elaboration

Level   Beginner +

Time   15 minutes

Materials

Two similar pictures with small differences.

Possible sources:

• a Photoshopped picture. See http://www.photoshop.com/ for a free, simple-to-use website that allows you to alter photos

• ELT resource books

• line drawings that you photocopy and, if you are of an artistic bent, alter yourself.

Preparation

1. Find or prepare two pictures: one original and one with eight to ten differences from the original. See references above for Photoshopping or use the sample visual.

2. Photocopy enough pictures so that every student has one or the other. In a class of 20, 10 students will have picture A and 10 will have picture B.

Procedure

1. (before listening) Put the students into pairs, A and B. Give the same picture to all the As. Give the Photoshopped version to all the Bs. Make sure As and Bs cannot see one another's pictures. You may wish to have them sit back-to-back or erect a barrier to prevent 'peeking'.

2. Explain that the students need to describe their pictures in order to find the differences (give the number) but cannot look at their partner's picture.

3. Elicit some useful phrases before starting, and write these on the board: at the bottom of my picture there's a . . . ; behind this there's a . . . ; did you say you have a . . . ?

4. (listening) Set a time limit and tell the students to begin.

5. (after listening) Once the time is up, ask which pair found all the differences. Then let the students see each other's pictures to check. (Optional: to inject grammar instruction you can ask students to state the differences explicitly.)

6. Go through the differences. This reinforces any difficult vocabulary.

Variation

Text differences. Student A reads her text (story) to Student B, while Student B identifies places in the text that differ. Students must identify a fixed number of differences, some of which may be at the level of grammar or lexis or discourse (e.g. order of events).

Comments

We have found that the activity is surprisingly rich in terms of language. Depending on the picture, it practises prepositions of place, there is/there are, follow-up questions, plus any vocabulary contained in the picture.

Research links

Batstone, R. (2002) 'Contexts of engagement: a discourse perspective on "intake" and "pushed output"', *System*, 30, pp. 1–14.

Charles Antaki's web page: 'An introduction to conversation analysis provides detailed guidance for transcription and analysis of natural conversations': http://www-staff.lboro.ac.uk/sscal/site

Creese, A. (2005) *Teacher Collaboration and Talk in Multilingual Classrooms.* Bristol: Multilingual Matters.

Lynch, T. (2009) *Teaching Second Language Listening.* Oxford: Oxford University Press.

Maleki, A. (2010) 'Techniques to teach communication strategies', *Journal of Language Teaching and Research*, 1, pp. 640–646.

Wright, A. (2003) *1000+ Pictures for Teachers to Copy.* London: Thomas Nelson Publishers. http://tell.fll.purdue.edu/JapanProj/FLClipart/.

Sample Visual

Source: Images reproduced from www.wisehat.com

Does your photo have a . . . ?

No, mine has a . . .

I've got a . . . in the left corner of my photo.

Me too.

Is there a . . . at the bottom?

No, I don't have one of those.

So that's one difference.

Activity: Whisper Dictation

Introduction

This activity deliberately creates difficult listening conditions in order to facilitate an awareness of the need for **collaborative strategies**. While most difficult listening conditions are cognitive (due to unfamiliarity with content or language), this activity creates difficult acoustic conditions. Because the students have to whisper, the speakers – as well as the listeners – need to use cooperative strategies, such as slowing down, repeating, confirming, rephrasing, modifying their input, emphasising certain words and articulating in exaggerated ways. The goal of successful interaction is achieved through the negotiation of meaning that comes after the dictation stage.

Aim
- develop the skill of questioning what you hear
- use strategies for negotiating meaning
- develop these active listening strategies: comprehension monitoring, persistent attention, contextual inferencing, seeking clarification

Level Beginner +

Time 15 minutes

Materials

Short passages or jokes.

Possible sources of jokes:

http://iteslj.org/c/jokes.html
http://www.1000ventures.com/fun/fun_ps_j.html1
http://www.lotsofjokes.com/

Preparation

Cut up short passages from magazines or newspapers so that each pair of students has their own passage or use the sample text in the worksheet.

Procedure

1. (before listening) Put the students into pairs, one scribe and one dictator. Give the dictator the short passage on a piece of paper. The scribe should be seated with pen and paper.

2. In a whisper, explain that they will be doing a whispering dictation, and only whispering is allowed today.

3. Move the dictators to the ends of the room so that they are far from their partner.

4. (listening) Explain that they have two minutes to dictate the passage, which the scribe will write down. (Two minutes is a guideline; this figure may vary, depending on the length of the passage and the students' level. The important thing is that it should be a difficult task, barely achievable in the time given.) Remind them to whisper.

5. After two minutes, tell the students they can now sit with their partners to check their texts, but that the scribe cannot look at the passage. He or she must ask questions only. Write example questions on the board: 'Can you repeat the first sentence?' 'What did you say after _____?'

6. (after listening) After a few minutes, tell the students they no longer have to whisper and get them to check their texts by looking at the passage.

Variation

The passages can be parts of a story. As a final stage, the students come together to work out the story.

Comments

Dictation is a rich activity type in that it involves the four skills (reading, writing, speaking and listening) and also there are numerous variations that can be used to provide different emphases. While this activity facilitates the use of repair strategies, other dictation types can help students practise grammar items, tricky areas of pronunciation, punctuation, vocabulary, etc.

With this activity we frequently see live evidence of students inventing and owning repair strategies. At first their main strategy is to simply ask, 'What?' But, after a while, they tend to expand their range of strategic gambits: 'Can you slow down?'; 'Wait a moment'; 'Hang on'; 'Say it again, please,' etc.

Research links

Davies, P. and Rinvolucri, M. (1989) *Dictation: New Methods, New Possibilities*. Cambridge: Cambridge University Press.

Nation, P. and Newton, J. (2009) *Teaching ESL/EFL Listening and Speaking*. New York: Routledge.

O'Malley, J. and Chamot, A. (1990) *Learning Strategies in Second Language Acquisition.* Cambridge: Cambridge University Press.

Oxford, R. (1990) *Language Learning Strategies: What Every Teacher Should Know.* Heinle ELT.

Wajnryb, R. (1990) *Grammar Dictation.* Oxford: Oxford University Press.

Wilson, JJ (2008) *How to Teach Listening.* Harlow: Pearson Education.

Worksheet

Audio File 12: Whisper Dictation

Please visit www.routledge.com/9781408296851 for an audio recording of this transcript.

Sample Text 1

Two women are very proud of their dogs. They begin discussing which dog is more intelligent.

The first woman says, 'My dog is so intelligent that every morning he waits for the newspaper boy to arrive, and then he puts the newspaper on the breakfast table.'

'I know,' says the second woman.

'How?' asks the first woman.

'My dog told me.'

Sample Text 2

A woman goes to a doctor, complaining of pain.

'Where does it hurt?' asks the doctor.

'Everywhere,' says the woman.

'Can you be more specific?'

So the woman touches her knee with her finger. 'Ow!' she says. Then she touches her nose, 'Ow!' Then she touches her back, 'Ow!' Finally, she touches her cheek, 'Ow!'

The doctor tells her to sit down, takes one look at her and says, 'You have a broken finger.'

Activity: Interrupted Story

Introduction

This activity helps empower students by demonstrating some of the vital sub-strategies of collaborative listening. The activity involves the students learning and trying out useful expressions for interrupting a speaker (asking for clarification, asking the speaker to slow down, etc.), which, in turn, allows them to gain some control over the input. Gaining control and developing some kind of give-and-take symmetry with the speaker is a vital part of active listening. In a way, it is the equivalent of giving them control of the pause, rewind and volume buttons on a DVD player.

Aim	• listen actively to a story or anecdote
	• develop these active listening strategies: backchannelling, problem evaluation, summarisation, double-check monitoring
Level	Beginners +
Time	20 minutes

Materials

A sequential story or anecdote.

Possible sources:

Fables: classic stories and fables, such as *The Stonecutter* (see the sample script)

http://www.rickwalton.com/pubtales.htm

http://www.youtube.com/watch?v=7ChOYWJV2U4

Stories with Holes: short mystery stories that require the listener/reader to ask yes or no questions to piece together the story. www.storieswithholes.com

Children's stories: free short illustrated stories, many with audio narration. www.magickeys.com

Phrases for interacting with the story-teller.

Asking for clarification:

• Wait. I didn't catch what you said.

• Hang on, what was that?

Asking for repetition:

- Sorry, can you repeat that?
- One moment. Can you say that again?

Asking the speaker to slow down:

- Hold on. Can you slow down a bit?
- Would you mind slowing down? I can't keep up.

Checking:

- So, what happened was that . . .
- So, have I got it right? The hero . . .

Preparation

Prepare to tell the story or anecdote.

Procedure

1. (before listening) Find a story that can be told chronologically (see the sample script). It could be a personal anecdote.

2. Elicit a number of phrases that can be used to interrupt a speaker: 'One moment'; 'Just a minute'; 'Sorry, can I interrupt you?'; 'Can I get something straight?'; 'Wait a second'. Also elicit expressions such as, 'Can you slow down?'; 'Can you repeat that, please?' Write the expressions on the board and drill them, paying particular attention to intonation.

3. Explain that you will tell a story but the students must interrupt you at appropriate moments. Their listening task is simply to understand the story.

4. (listening) Begin telling your story. At certain points, speed up, speak too quietly, and use some above-level vocabulary. The students call out the expressions, as appropriate.

5. Explain that you will retell the story. The students are to listen to check their understanding only. Tell them that *they* will retell the story in pairs afterwards. Begin retelling.

6. (after listening) Before the students retell the story, remind them of the expressions on the board. While working in pairs, the students may use these again, where appropriate.

Variation

Tell a story badly. After teaching the expressions on the board, discuss what makes a good storyteller. Elicit ideas such as: the story is told in a clear voice and the story moves from one point to the next in sequence. Explain that you will tell a story badly. The students must use the expressions on the board to help you improve. (Please slow down. Please speak more clearly. Please connect your ideas better . . .) As you tell the story, it needs to be exaggeratedly, comically bad – in other words, *unacceptably* bad. This forces the students to intervene. When done skillfully, the badly told story can produce much laughter in class, as well as clear opportunities to use the target expressions.

After this initial listening activity, retell the story well. The students should hear a good version in order to enjoy the story, settle any doubts about content, and provide a model for their own storytelling.

Comments

We have found it useful to emphasise to the students that they will need to retell the story. This provides a different motivation to listen carefully the second time.

Research links

Block, D. (2003) *The Social Turn in Second Language Acquisition*. Edinburgh: Edinburgh University Press.

Bremer, K., Roberts, C., Vasseur, M., Simonot, M. and Broeder, P. (1996) *Achieving Understanding*. London: Longman.

http://www.teachingenglish.org.uk/articles/story-telling-language-teachers-oldest-technique.

http://www.creativekeys.net/storytellingpower/article1004.html.

A lecture by Brian Sturm on storytelling theory:
http://www.youtube.com/watch?v=UFC-URW6wkU&feature=related.

Sample Script

Audio File 13: Interrupted Story

Please visit www.routledge.com/9781408296851 for an audio recording of this transcript.

Upper Intermediate/Advanced

One day, a poor stonecutter worked chipping stone from the side of a mountain. Tired and hungry, he said, 'I wish I'd been born a rich man.' A magic spirit heard him and transformed him into a rich man. He enjoyed his new life, but one day the sun burned him. He said, 'The sun is more powerful than me. I wish I was the sun instead of a rich man.' So the magical spirit transformed him into the sun. Now he shone down on the earth. But one day a cloud passed in front of him. 'The cloud is more powerful than me! If I'd known this, I would have asked to become a cloud!' The magical spirit turned him into a cloud. Now he blocked the sun and caused cold weather, but one day the wind blew the cloud away. He said, 'If I'd been stronger, I could have stopped the wind. I wish I was the wind.' The spirit granted his wish. He blew and blew, creating dust storms and hurricanes, but when he tried to blow a mountain over, he failed. 'If only I was stronger, I would have blown that mountain down. I wish I was a mountain.' Again, the spirit helped him. Now, he stood huge and immovable. But one day he felt something chipping at him. It was a stonecutter. 'The stonecutter is the strongest of all!' he said. 'If only I had known this, I would have remained a stonecutter. I regret changing my life, and I want to be a stonecutter again.' And so the magic spirit turned him back into a poor stonecutter.

Activity: Interactive Quiz

Introduction

Structured teacher–student interaction has long been a staple of both inter-action and assessment in the classroom. The main problem with the frequent question-response-comment (Teacher: What's the largest city in Germany? Student: Berlin. Teacher: Good.) format of typical classroom discourse is that students come to believe this 'permission pattern' is the only way they can use the L2. This activity extends the traditional practice of calling on students to give short responses to factual questions or opinion questions. While still in a test-like format, this activity gives students a chance to *anticipate a range of questions* they will encounter and to *stay in contact with the speaker* to keep the conversation flowing smoothly.

Aim	• lessen anxiety about oral interaction and oral assessment
	• promote interaction with the teacher
	• create homework for conversation classes
	• develop these active listening strategies: advance organising, backchannelling, linguistic inference
Level	Beginner
Time	2 minutes per student

Materials

Thematic lists of questions or prompts given to students in advance (see sample quizzes).

Practice tests for speaking exams, such as the IELTS speaking exam, will provide conversational themes and questions that you can adapt for this type of interactive quiz:

http://www.goodluckielts.com/IELTS-speaking-topics.html

Example, weather

- Does the weather affect your mood?
- How do rainy days make you feel?
- What's your favourite season of the year?
- What do you like to do when it's hot?
- What do you usually do in the winter?

Preparation

Give the students a list of topics, questions or question types you will ask *in the next class*. The list should include more topics and questions than you will actually ask, so that the students are preparing, but not memorising fixed responses. You can dictate the list of topics or questions, write it on the board, or put it on a class website.

Procedure

1. (before listening) Announce that you're going to start the day's speaking quiz: 'OK, ready for today's speaking quiz?'

2. (listening) Start by asking a question or giving a prompt (it's best to do this randomly rather than in a fixed order, as it improves memory and promotes intensive listening and quick thinking): for example, 'Tell me about your weekend. What did you do?'

3. Call on a student to answer. (This is preferable to calling on a student first, as it keeps everyone listening more carefully.)

4. Continue until you've gone through the list of topics or questions or have talked with a set number of students.

5. (after listening) Provide some global feedback to the group, about grammar, vocabulary, pronunciation, or pragmatic communication strategies (such as how to use **fillers**, like 'Umm . . .' or 'Let me think') to maintain contact with the speaker while you're thinking of how to answer.)

6. Have the students work in pairs to practise asking and answering the same questions you used in the day's interactive quiz.

Variation

Scored quiz. Use this type of structured interaction as a scorable speaking quiz. After the student answers, respond by saying, 'Good' or, if there's a small error, say something like, 'Oops, there's a small error.' See if the student can correct the error. If you don't understand the response at all, say, 'Sorry, I don't understand', and have the student try to reformulate. Record a score for each student, using a simple rubric, or a 1–5 scale, with a 5 being a perfect response for a student at this level.

5 = Perfect

(Students give a quick communicative answer with few or no errors in pronunciation or grammar.)

4 = Small mistake(s) (Close!)

(Understandable, but with some small errors in pronunciation or grammar.)

3 = Big mistake(s) (Nice try!)

(Way off, but made an attempt.)

2 = 'Sorry, I don't know.' (Escape!)

1 = If you say nothing (after 5 seconds)

Comments

We've found that this kind of structured speaking activity can be very useful for beginner classes that are reticent to speak and, especially, for large classes. If you have a regular speaking-listening activity like this to begin a class, it can help students feel involved at the outset and inclined to participate more actively in class. You can create interactive quizzes that are very personalised for your particular class, talking about the students' interests and activities.

Research links

ESL go website has grading rubric for speaking tests:
http://www.eslgo.com/resources/sa/oral_evaluation.html

Joshua Kurzweil's article on interactive quizzes:
http://iteslj.org/Techniques/Kurzweil-OralQuizzing.html (*The Internet TESL Journal*, 11)

Sample Quizzes

Beginners: Tell the students in advance they should be prepared to see these actions in the quiz.

Quiz for _____ (date): Practise the past tense

Example

Teacher: (Gestures: raising hand) What did I do?
Student: You (rais<u>ed</u> your hand).

Be prepared for these actions in the quiz:

point to	eat	sing (a song)
scratch	drink	hum (a tune)
rub	stand on	kick
tap	sit on	bump into
walk toward	jump over	write
try	hop	touch
turn	smile (at you)	frown
put (something on something)		

Advanced: Tell the students in advance they should be prepared to answer questions on these topics.

Travel

- Do you like to travel?

- What kind of places have you visited in your life?

- Which place would you really like to visit? Why?

- What's the best place you've ever visited?

Computers

- Do you think computers help society?

- Do you think computers are bad for health?

- How do you think computers have changed the world?

Activity: Blind Forgery

Introduction

This task is a multi-modal listening activity, enriched by the use of authentic art from different world cultures. The activity is particularly enjoyable for design-oriented students who respond well to visual images. The listening aspect of the activity is made interactive through the need for constant negotiation of meaning. To complete the task, the 'artist/listener' needs to understand what the speaker really means. The addition of time pressure also gives the activity a kind of game-like feel, which many students enjoy.

Aim
- listen for specific detail
- develop the skill of negotiating meaning
- develop these active listening strategies: advance organising, self-management, collaborating, joint task construction, seeking clarification

Level Intermediate +

Time 20 minutes

Materials

You will need one fairly simple picture, without too much detail, per student. These could be famous artworks, photos from magazines, faces, landscapes, rooms, etc.

Possible source:

Google Images: http://www.google.com/imghp

Preparation

1. Find pictures suitable for students to describe and draw.
2. Get the pictures into an appropriate form for the students to hold and describe (e.g. cut out and laminated).

Procedure

1. (before listening) Put the students in pairs. Give each student a picture, to be concealed from their partner.

(Option) Half the class can have one picture while the other half has a second picture.

2. The non-speaking student has a blank piece of paper on which to draw. Explain that the 'speaker' must describe the picture so their partner can draw it.

3. Explain that the 'listener' can ask clarification questions. Elicit some and write them on the board: 'Is it big or small?' 'What shape is it?' 'Does it go behind or in front?' 'Can you repeat that, please?' 'What exactly does it look like?', etc.

4. (listening) Set a time limit (probably not less than four minutes) depending on the level of the students and the complexity of the pictures. Be flexible with this time limit. The students describe and draw the pictures.

5. The artist shows the talker the picture he or she has drawn. The talker corrects it by asking the artist to redraw parts.

6. When time is up, give the 'listeners' a chance to look at the original picture.

7. The students change roles and follow stages 2–4.

8. (after listening) Display the originals and the copies on the walls. The students wander round the room as if at an art gallery.

9. (Optional) Add a blank piece of paper next to the pairs (the original and the copy) and ask students to write brief comments about their classmates' efforts as they circulate.

Variation

My room. Have one student describe their room at home to their partner. Include major objects in the room and their location. The listener should attempt to control the activity by asking questions: Do you have a . . . ? Where is . . . ? How big is . . . ? and by showing the illustration in process: Did I get this right?

Comments

We have found that the activity works particularly well with paintings by a number of figurative artists, e.g. Gauguin, Van Gogh, Cezanne, Edward Hopper, Marc Chagall, as well as regional/local art and so-called 'Primitive Art'.

We have used this activity to help students rehearse for parts of certain exams, e.g. Cambridge First Certificate, which require students to describe and comment on pictures.

Research links

Goldstein, B. (2009) *Working with Images*. Cambridge: Cambridge University Press.

Keddie, J. (2009) *Images – Resource Books for Teachers*. Oxford: Oxford University Press.

Sample Visual

The Card Players by Cézanne
Courtesy of Bridgeman Art Library

Activity: Milestones

Introduction

Like other interactive listening activities, Milestones involves speaking. Indeed, this particular activity is highly speaker-focused in that it is the speaker who determines the content of the stories. However, the activity works only if the listeners become very curious and begin probing for more information. The use of the Post-it® Notes or cards with dates serve as a kind of tangible reference and planning device that helps keep information in the conversation active. In this way, the students can jump back and forth between topics, as we do in normal conversations.

Aim
- listen for details of personal stories
- develop the strategy of questioning to keep a conversation going
- develop these active listening strategies: advance organising, double-check monitoring, directed attention, seeking clarification

Level   Intermediate +

Time   45 minutes

Materials

Six blank Post-it® Notes (or blank cards or small pieces of paper) per student (see sample visual).

Preparation

1. Prepare to spend a few minutes talking about two or three milestones from your life.
2. Have six Post-it® Notes per student, ready to hand out.

Procedure

1. (before listening) Elicit some important events in a person's life, e.g. graduated, met partner, travelled abroad, got a job, got married, etc. Explain that the students will be talking about these. Write two or three dates on the board and ask the students to guess which important events occurred for you on these dates. Say what happened, describing each event in about one minute.
2. Tell the students they're going to talk about six events/milestones in their lives. Give them a few minutes' thinking time.

3. Hand out six Post-it® Notes (or blank cards or small pieces of paper) to each student, and tell them to write the year when the events took place – one year on each note. Actually, it's best to *show* the students they need to write one date on each note, as invariably, one or two students write all six dates on one!)

4. Put the students into groups of four. Ask them to arrange *all of their dates combined* into chronological order in a circle (i.e. all 24 dates). Monitor as they do this. The circle should look like a clock with 24 numbers (see sample visual).

5. (listening) Explain that, as a group, they will go around the circle and whoever's Post-it® Note it is must talk for one minute (or 30 seconds for lower levels) about the event that took place during that year. The others must keep the student talking by asking questions.

   Example:
   A: 2009 is important to me because it's the year I met my girlfriend.
   B: What's her name?
   A: Nobantu.
   C: Where did you meet her?
   A: At a party.

6. As the students move around the circle, monitor and take notes for a feedback stage.

7. (after listening) At the end, ask the students to say one thing about each member of their group that they learned today.

8. Provide a brief feedback session, pointing out any useful expressions they used and highlighting some common errors.

Comments

We have used this activity with all levels from low intermediate upwards. At lower levels the focus is on using the past simple tense and developing the conversational strategies of showing interest and probing for more information. At higher levels, the focus is on using a variety of tenses (e.g. past perfect, past perfect continuous) and rich idiomatic language to describe feelings, as well as the communication strategies.

Research link

Griffiths, G. and Keohane, K. (2000) *Personalising Language Learning.* Cambridge: Cambridge University Press.

Sample Visual

Time Circle

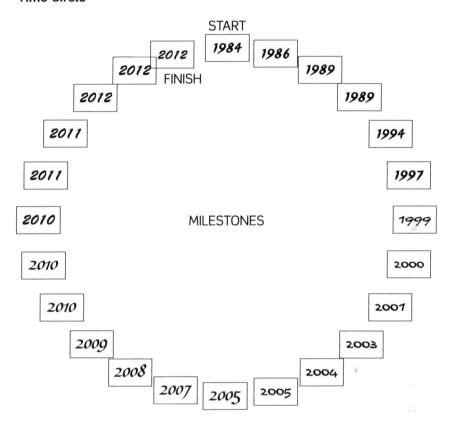

Activity: My Turn/Your Turn

Introduction

To develop the habit of active listening, it is useful to practise in both game-like and more academic contexts, with both light and rich content. This activity involves students sharing responses to any piece of content-rich input. Particularly for longer listening passages or detailed stories, the collaborative aspect can help students to piece together the story while checking what they heard. The activity involves two types of listening – listening to the passage, then listening to your partner's interpretation of events from the passage. This type of jigsaw activity helps build up the learners' attention span.

Aim
- focus on discreet units of information while listening
- develop post-listening skills of paraphrasing and checking information
- develop these active listening strategies: advance organising, question elaboration, joint task construction

Level Intermediate +

Time 10 minutes

Materials

Audio extracts up to five minutes in length

Possible sources:

http://storycorps.org/listen/

http://www.bbc.co.uk/programmes/p00ms6zb

http://www.npr.org/

http://yourstorypodcast.com/ (podcasts of people telling interesting stories about their lives)

Preparation

Check the recording and provide some gist questions for the first listening (see sample script).

Procedure

1. (before listening) Prepare students for listening by discussing the topic, showing a picture, asking questions, etc. Tell them they will listen to the whole recording twice.

2. Set a general gist question or two for the first listen-through.

3. (listening) Play the recording and answer the question(s) in plenary.

4. Tell the students they will hear the passage again and must take turns afterwards to explain any pieces of information that they remember. Give an example, focusing on one unit of information from the beginning of the passage. Remind the students not to take notes, just to pay attention to the information in the passage.

5. Play the recording.

6. (after listening) Put the students into pairs and have them take turns to say one unit of information each.

7. Go round the class eliciting the information they mentioned.

Variation

*   **Group summary**. Put the students into groups of four and have them write a summary of the passage. The groups compare summaries in order to see what information they omitted or whether their interpretation accords with others'.

*   Get the students to work in pairs and choose their own input (e.g. BBC news or NPR hourly news summary) before doing stage 6 of My Turn/Your Turn.

*   **Dictogloss**. After hearing the passage a number of times, the students work in pairs or groups to produce a written version of the text. The emphasis is on producing a workable version of the text rather than an exact transcription. The teacher can exploit the student-produced texts for grammar and vocabulary instruction.

Comments

We use My Turn/Your Turn particularly with listening passages that contain a lot of information to hold in the memory. The sharing of the burden of remembering builds confidence in the students and develops a collaborative approach to oral text reconstruction. It also means that the shyer students get their chance to speak in a non-threatening context, where they will not be cowed by more confident students dominating the conversation.

The type of factual story appropriate for this activity is common in Part II of the Listening Paper of the Cambridge Advanced Exam (CAE), which consists of a factual monologue, and Section 4 of the Listening Paper in the IELTS exam.

Research links

Jacobs, G. and McCafferty, S. (2006) 'Connections between cooperative learning and second language learning and teaching' in McCafferty, S., Jacobs, G. and Iddings, A. (eds) *Cooperative Learning and Second Language Teaching*. Cambridge: Cambridge University Press.

McCafferty, S., Jacobs, G. and Iddings, A. (2006) *Cooperative Learning and Second Language Teaching*. Cambridge: Cambridge University Press.

Nunan, D. (1992) *Collaborative Language Learning and Teaching*. Cambridge: Cambridge University Press.

VanPatten, W., Inclezan, D., Salazar, H. and Farley, A. (2009) 'Processing instruction and dictogloss', *Foreign Language Annals*, 42, pp. 557–575.

Sample Script

Photo courtesy of Getty Images

Questions (Advanced level)

1. Who were John and Washington Roebling?
2. What was remarkable about the building of the Brooklyn Bridge?

Audio File 14: My Turn/Your Turn

Please visit www.routledge.com/9781408296851 for an audio recording of this transcript.

How the Brooklyn Bridge Was Built

Connecting Manhattan Island to Brooklyn, New York, there is a very famous structure called the Brooklyn Bridge. It's a remarkable piece of engineering and it comes with a remarkable story. In 1863, an engineer called John Roebling devised a plan for the bridge. His contemporaries, however, didn't believe it would be possible. The distances were too great and there were too many obstacles.

But Roebling refused to give up. He discussed the idea with his son, Washington, who was also an engineer, and the two of them worked out how to tackle each problem in turn. Eventually, they hired a crew and began work on the bridge.

Just a few months after they'd started, a tragic accident occurred at the site, killing John Roebling. His son Washington took over, but in the early 1870s, while at the site, he too became involved in an accident from which he never recovered. With his condition degenerating, he was unable to visit the site. No one believed the project would continue as Roebling Senior and his son had been the driving forces behind the construction of the bridge and no one else knew exactly how they were planning to build it.

However, everyone had underestimated Washington Roebling and his determination to complete the project. Perhaps spurred on by the memory of his father, Washington decided to continue as chief engineer, proceeding by giving instructions to his wife on how to complete the bridge. She then passed on his messages to the engineers.

Using this slow, laborious process, Washington continued giving instructions for 13 years. Finally the bridge was built, a truly remarkable achievement against almost impossible odds.

Activity: Guest Speaker

Introduction

Teachers often worry that their students have become accustomed to their teaching and language styles, and that these students are not transferring their listening skills to English spoken outside their class. One way to counterbalance this tendency is to bring in guest speakers. Guest speakers need not be celebrities, and they need not be native speakers. In fact, it is often very motivating for students to have a non-native guest speaker. Also, guest speaker spots can be as short as 5 or 10 minutes. This activity outlines a framework for ensuring that your guest speaker experience is successful for both your students and the speaker.

Aim
- practise listening and interaction skills in a new context
- develop these active listening strategies: advance planning, predictive inferencing, world elaboration, resourcing

Level Intermediate +

Time 30 minutes

Materials

(Optional) A pre-reading that gives the student some background about the speaker or the topic.

A worksheet to guide students during the session.

Preparation

1. Brainstorm the local speakers you could invite into your classroom to give short talks on various topics of interest. If you are not in an English-speaking country, you may still be able to find speakers from cultural organisations, among the expatriate community, or from among your friends and colleagues. And remember that guest speakers need not be native speakers, but they should be reasonably fluent and comprehensible.

2. (Optional) Select or create a short reading that focuses on the topic that the speaker will discuss, and that includes a number of key concepts and vocabulary items likely to occur in the talk.

3. Talk to the guest speaker about the length and format of the talk. It's a good idea to give speakers some tips for organising their presentations and for interacting with the students.

Procedure

1. (before listening) (Optional) Before the speaker arrives, have the students read and summarise the extract you have selected. Based on the reading, you may ask the students to prepare some questions for the speaker. (Sometimes it is nice to have a surprise guest speaker and *not* have any preparatory activities!) Alternatively, distribute a guest speaker summary worksheet (see sample).

2. (Optional) Ask one student to be responsible for introducing the speaker briefly and for thanking the speaker. If possible, the student should chat with, or at least email, the speaker prior to the talk to get some biographical information.

3. (listening) For the guest presentation, organise the session so that there is time for questions after the talk.

4. (after listening) Have a formal debriefing session with the students after the speaker has left: Did they enjoy the guest speaker? What was the most enjoyable part for them? What was most challenging? Did they understand most of what the speaker said? Did they notice the differences in language and style that the speaker used and the language and style you use?

5. Pose three or four open-ended questions that the speaker touched upon in the talk. Have the students discuss these in small groups.

Variation

Virtual guest speaker. Although it is not as compelling as a real live presentation, you can have a virtual guest speaker join you via a web cam. For this format, it is enjoyable to have a surprise guest. The students can still interact via questions, and you can still have a follow-up debriefing and assignment.

Comments

When we have taught in EFL contexts, we have readily found a number of volunteer English-speaking experts (not necessarily native speakers) to donate an hour of their time to speak to the students about what they do professionally, or about a particular hobby, interest or belief they have. Volunteers have included Olympic athletes, world travellers, artists, nutritionists, beekeepers, magicians and missionaries (who agreed not to proselytise). We generally encourage the speakers to bring a few visual aids, but to avoid detailed PowerPoint presentations.

Using a worksheet will help students follow the talk, formulate questions, and later summarise. You may wish to create a customised worksheet after you have learned the intended structure of the speaker's talk.

Research links

BusyTeacher website has ideas for getting the most from a guest speaker: http://busyteacher.org/7083-top-10-ways-get-most-from-guest-speaker. html.

Koester, A. (2010) *Workplace Discourse*. London: Continuum.

Ross, S. and Berwick, R. (1992) 'The discourse of accommodation in oral proficiency interviews', *Studies in Second Language Acquisition*, 14, pp. 159–176.

Worksheet

Guest Speaker Summary

Name of the speaker

Occupation

Some biographical information

What is one specific accomplishment of the speaker?

How would you describe this speaker?

What questions would you like to ask the speaker?

Activity: Paraphrase

Introduction

The main aspect of active listening in interaction is creating alignment with the speaker. Alignment can be initiated through backchannelling – showing the speaker that you are involved in the interaction. Backchannelling relates to several aspects of active listening, including showing that you are following the speaker's meaning and are ready to continue the conversation. This activity provides opportunities for students to develop the strategy of paraphrasing, which is perhaps the most challenging form of backchannelling. The activity provides some sample expressions for soliciting opinions and paraphrasing in socially sensitive ways.

Aim
- improve symmetry in conversations
- elicit opinions and clarifications from others
- practise paraphrasing
- develop these active listening strategies: seeking clarification, summarising, mediating

Level Intermediate+

Time 20 minutes

Materials

A list of current topics, including controversial ones, on which students will have clear opinions. See the worksheet or come up with your own topics.

Sources:

http://www.buzzle.com/articles/controversial-topics-for-research-paper.html

http://www.speech-topics-help.com/controversial-speech-topics.html

Preparation

Select 10 topics and formulate them as statements that elicit an 'I agree' or 'I disagree' response. (See the worksheet.)

Procedure

1. (before listening) Review useful language for soliciting opinions, express-ing opinions and paraphrasing (see sample phrases).

2. Pass out the worksheets. Ask students to read the statements and decide if they basically agree or disagree. Or, to allow for shadings of opinion, you can use a Likert scale: 1 = completely disagree to 5 = completely agree.

3. (Optional) Have students work in pairs to give their opinion, and then to give one 'pro' and one 'con' reason in support of, or in opposition to, each statement.

4. (listening) After you have completed the list, ask the students to stand up and mingle. They should exchange opinions about one of the topics with one or two classmates. Note: the focus is on understanding their partner's opinion, not on agreeing or disagreeing, and not on trying to change their partner's opinion.

5. Each listener must paraphrase their partner's opinion before giving their own opinion. Use the worksheet for examples. After exchanging opinions on one topic, students should find a new partner and continue with a different topic.

6. (after listening) At the end, as a whole class, summarise the range of opinions about one or two topics. (Optional: parts of this summarisation can be in the students' L1; work on translating the L1 opinions into colloquial English.)

7. As a follow up, have the students choose one topic for research, and get them to write a short 'position paper' on the topic. The students should aim to have three supporting reasons for their position.

Variation

Debate. Divide the class into teams, one for and one against a proposition. Give them time to list arguments in support of their point of view. Have an exchange of views between the teams, one member at a time. Allow a fixed time for questions about each argument. At the end of three to five turns, decide which side 'wins' the debate – not who has the 'best' opinion, but who expressed and defended their views most thoroughly.

Comments

We have often heard students say how much they enjoy debates in language class – even though debates usually require more work than other activities.

Some students have said that debates help improve their ability to think in English – and make other activities seem simpler in comparison. Debates also force comprehensible output, the need to make your views explicit and understandable to others.

Research links

Bardovi-Harlig, K. (2006) 'On the role of formulas in the acquisition of L2 pragmatics' in Bardovi-Harlig, K., Felix-Breasdefer, A. and Omar A. (eds) *Pragmatics and Language Learning*, Volume 11. Honolulu: NFLRC.

Imai, Y. (2010) 'Emotions in SLA: new insights from collaborative learning for an EFL classroom', *Modern Language Journal*, 94, pp. 278–292.

Wolf, J. (2008) 'The effects of backchannels on fluency in L2 oral task production', *System*, 36, pp. 279–294.

Worksheet

Prepare a for/against sheet (enlarge as necessary)

	agree (+) or disagree (–)	one argument for	one argument against
Smoking should be banned in public places			
The internet should be censored			
Schools should ban the sale of junk food on campus			
Medical testing on animals should be stopped			
Corporal punishment of children should be illegal			
Homework should be limited to one hour per day			
Schools should provide for single sex instruction			
Zoos do more harm than good			
Children under the age of 18 should not be allowed to use Facebook			
Children should not be adopted by same sex couples			
Students should be required to wear uniforms to school			
(your idea)			

Sample Phrases

Soliciting

Can you tell me your opinion about . . . ?

What's your view on . . . ?

On the issue of . . . , are you for or against?

I wonder if you could give me your take on . . . ?

How about you? What do you think?

Paraphrasing

So you're saying that . . . ?

If I understand correctly, you mean that . . . ?

Let me see if I got this right. You believe that . . . ?

Are you saying that . . . ?

In other words, it seems you're saying that . . . ?

Is that right? / Did I get that right?

Activity: Pecha-Kucha

Introduction

Presentations are often seen as essentially one-way communication, with the audience having few, if any, opportunities to evaluate, provide feedback, or interact with the speaker's content. For this reason, the students who are the listeners (the majority of the class) often find presentations to be boring and unproductive. This activity provides opportunities for speakers and listeners to interact more, for the speaker to become more sensitive to the audience, for the listeners to check their comprehension of the speaker's ideas and to give an evaluation of the speaker's impact on them. Keeping the audience involved in ways like these is essential for teaching two-way communication.

Aim   • practise supportive listening

- practise asking questions to presenters

- practise reconstructing the speaker's content

- develop these active listening strategies: advance organising, revision evaluation, seeking clarification, summarisation

Level   Intermediate +

Time   5 minutes for each presentation (with audience recreation step)

Materials

Computer and projector for student presentations; software should have a timer for timing each slide (or a fellow student can forward slides using a stopwatch).

Preparation

1. Tell students to prepare a presentation on a given theme for a future class. Each presentation should have a set number of slides (from 5 to 10). Tell the students they will have 20 seconds (only! no more, no less) to talk about each slide.

2. (Optional) Show some sample **Pecha-Kucha** style presentations as examples: http://www.pecha-kucha.org/presentations/ (Example of a non-native speaker giving a presentation in English on knitting http://

www.pecha-kucha.org/presentations/289.) You can also have students fill out the evaluation form on the worksheet for a sample presentation to get used to the descriptors.

3. Give students time to rehearse their presentations in pairs or small groups outside class, or in class (using a mobile phone, if available, to show their slides).

Procedure

1. (before listening) Obtain the slides that each presenter is going to use. Put them into a PowerPoint or similar software program, with the duration of each slide set for 20 seconds.

2. (listening) For the presentations, have the presenter stand in front of the class. You or a monitor should control the slide show, projecting each image for the agreed amount of time. It's OK for students to be silent for some of the time, but flip to the next slide, even if a student is still talking about the previous one – this keeps the procedure fair for everyone.

3. (after listening) At the end of the presentation, the audience should show their appreciation (applause!). The speaker should absorb the appreciation (bow, etc.). It's important for students to feel this acknowledgement for their efforts, even if they've experienced language or technical problems.

4. Give the class one or two minutes to ask the presenter questions.

5. Replay the slides. This time, the listeners, working in pairs, see the slides again with the same timing and have to reconstruct the talk. The original presenter can circulate and listen to what the audience is saying.

6. Ask students to fill out an evaluation form for the presenter. The presenter and the instructor should have the opportunity to read the forms.

Variation

Group presentations. Use the same slide-based set up, but have students work in pairs or groups. They alternate turns to talk about the slides.

Comments

We first learned about Pecha-Kucha through the architectural firm Klein-Dytham in Tokyo, which is the group credited with originating this pop phenomenon. The company developed this format so that younger, less experienced architects and architecture students would have 'equal time' in meetings and conferences that are usually dominated by more seasoned colleagues. It turned out that all participants liked the idea of 'levelling the playing field', a communication phenomenon that has spread widely.

Research links

http://academics.smcvt.edu/cbauer-ramazani/cbr/iep/spkg.htm.

http://teachesltech.vfowler.com/2010/12/pecha-kucha-presenting/.

Worksheet

Presenter Evaluation. At the end of each presentation, fill in this form

Speaker's name

The talk was about

It was easy/difficult to follow because

One thing I learned was

Feedback to the speaker

- *You were (not so/a little/very) enthusiastic*
- *Your voice was (not so/mainly/very) clear*
- *You interacted (a little/a lot) with the audience*
- *I liked the part where you said*

  _____

  _____

- *My advice to you would be*

  _____

  _____

Frame 5
Autonomous Frame

Introduction

The **Autonomous Frame** deals with independent listening activities, activities the learners can do on their own outside the classroom. This frame is concerned with locating suitable resources and *taking advantage* of those resources in ways that sustain motivation and support acquisition of listening at progressively higher levels. Autonomous listening hinges on finding ways to access and learn from authentic sources and developing strategies for dealing with challenging input.

This frame addresses questions such as: How can I motivate my students to listen outside class? Is there a way to bridge in-class and out-of-class learning? What sources are best for my students? How can I get my students to become independent? How can I get my students to become more *inter*dependent? Can I help them develop strategies to *enjoy* listening outside of class? What can I do if students feel overwhelmed by the difficulty of the language they're encountering? What kind of feedback do I need to give students?

Case studies of successful language learners consistently show that the most successful learners are able to learn on their own. They know that the classroom alone is not sufficient to provide the resources and opportunities they need to develop continuously and to a high level. Successful learners are seekers: they find ways to learn more on their own. The activities in this section aim to support them in that quest. For all activities in this frame, it is essential to have learners focus on developing strategies of planning, focusing attention, evaluating and reviewing. (See Appendix 1 for a list of active listening strategies and examples of specific strategies the students can try out.)

Ten illustrative activities

The ten activities presented in the Autonomous Frame are:

Activity	Description	Adding out-of-class learning opportunities	Connecting to social contexts	Integrating new technologies	Instructor support	Developing learning strategies
Transcripts	working with audio scripts and subtitles	■	■	■		■
Cloud Discussions	using online platforms to interact with peers	■	■	■		
Listening Games	using web resources/apps for listening practice	■		■		
News Hound	summarising news stories	■	■	■		
Vox Pops	talking to English speakers outside class		■			■
Webquest	doing an interactive research project	■		■		■
My Listening Library	developing a bank of useful resources	■		■		
Learn Something New	structuring and sharing new learning	■	■	■		■
Film Review	sharing and comparing film reviews			■	■	■
Conversation Corner	starting and maintaining a chat centre	■			■	

(Darkened boxes indicate research links to the activities.)

Activity: Transcripts

Introduction

Transcripts and subtitles are great resources for language learners. The main benefit is that, unlike sound, which disappears on the wind, transcripts allow the listener to re-read and check what they heard, and to really notice new language. By comparing the input in two channels, the listener gets to notice the differences between the way words are written and the way they are pronounced in context. As permanent records of evanescent speech, transcripts also provide a rich resource available for further study; students can 'mine' transcripts for useful vocabulary, discourse markers, grammatical features, etc.

Aim   • use a transcript or subtitles to aid listening

　　　　• develop these active listening strategies: comprehension monitoring, double-check monitoring, persistent attention, resourcing

Level   Beginner +

Time   20 minutes

Materials

Possible sources:

Most course books contain the transcripts of at least some of the recordings in the back of the book.

Course books with filmed material often include subtitles that can be switched on and off by the user. See *Speakout* (Pearson, 2011) for an example of authentic film material and specially filmed clips with a subtitle option.

Many commercial DVDs now include subtitles.

Preparation

Review the transcript while listening to the recording.

Procedure

1. (listening) The students listen to a recording or watch part of a film without subtitles. Their first task is comprehension. If using the worksheet, they can write a one-line summary of the passage.

2. They listen/watch again and note down words and phrases as they hear them, using the pause button, if necessary.

3. They listen/watch a third time, but this time with the subtitles or transcript in front of them. As they read and listen, they confirm or disconfirm the notes they took during the second listening.

4. (after listening) (Optional) The students prepare to summarise what they saw/heard to their classmates, and include three new words/expressions that they learned while listening/viewing independently.

Variation

- **From listening to speaking.** Ask the students to circle a few expressions they'd like to use in their own speaking output. Get them to note these down in their vocabulary books and write personalised example sentences which they share with other students.

- **Gapped transcripts.** This is the classic cloze test that appears in many exams such as CAE. Using a photocopy of the transcript, white out a number of key words and get the students to fill in the gaps as they listen. They may do predictive work (filling in the gaps before listening) first, then listen to check their answers. This activity combines aspects of bottom up and top down processes.

- Students can be asked to make their own transcripts of short clips or longer extracts. They then compare their transcripts with what their classmates wrote. This tends to highlight areas of difficulty in terms of listening, and serves as a useful noticing activity.

Comments

We have found that some teachers are resistant to using transcripts and subtitles because this sometimes leads to 'cued reading' rather than listening. Perhaps a solution is to use transcripts and subtitles only in a checking phase and only after the main work of comprehension has been completed. Our own students frequently come to class with transcripts that they have annotated at home, and wish to know urgently what this word or that expression means. They see the transcript as a tool for focusing on new 'heard' language.

Research links

Wilson, JJ (2008) *How to Teach Listening.* Harlow: Pearson Education.

Worksheet

Title/Topic

Short summary

Words and expressions

New words and expressions

Activity: Cloud Discussions

Introduction

While in class, teachers may be able to create valuable student–student interaction and discussions. But teachers wonder how to extend this aspect of oral language development outside of class meeting times. This activity outlines a method for using free online software to generate student oral presentations and discussions, as well as meaningful peer-to-peer listening practice. Using new technologies in this way can be very motivating for students, and can really boost learning opportunities. If the teacher can link out-of-class discussions to in-class content, there will be even more value.

Aim	• interact with other students outside of class time
	• get extra speaking and listening practice
	• develop these active listening strategies: self-management, personal elaboration, performance evaluation
Level	Beginner +
Time	20 minutes to set up

Materials

Websites that allow for free voice-to-voice interaction:

http://www.voicethread.com

http://www.lpadio.com

http://www.skype.com, including audio recording device (such as Ecamm, Soundflower)

Preparation

Visit http://www.voicethread.com for a demonstration of a 'cloud-based' activity that allows you to upload different kinds of media, such as images, videos, documents and presentations into a single place to enable an asynchronous (not in real time) group conversation. Once you upload the media, it works like a PowerPoint-style slide show. Any student who has a link to the presentation can make comments on the presentation. They can make comments via their computer microphone, webcam, text, or by uploading an audio file. These comments then appear in the 'voice chain', so that everyone in the class can review the comments as well as see who produced them.

Procedure

1. (before listening) Upload a simple speaking task to your voicethread page. (See the Sample Screenshot for an example.) Assign the students a time limit. For example, read the discussion question. Give your response in 30–60 seconds.

2. Demonstrate to your students how to log in to the class page.

3. (listening) Tell them to record and post their response. (VoiceThread provides the speaker with a chance to review what they've recorded before posting, and to revise as often as they wish.) Give a model, either in class or on the VoiceThread page.

4. Tell the students to comment on at least two other student comments as well. (For example, here is my comment on Jamal's answer: . . .)

Variation

Facebook photos. Instead of using online resources, you can have students access copies of photos or videos, and record their comments on a handheld device, or email the audio files to you for review.

Comments

We have found that including this type of speaking/listening activity in a course can be very motivating for students. Activities like this can be useful preparation for the interview portion of speaking tests such as IELTS, in which the test-taker has to speak for up to a minute on a specific topic. Because the students' contributions are archived on the Voicethread page, you can use these for speaking assessments.

The Sample Screenshot shows an example of a VoiceThread page. Participants post audio, video or text responses to a topic. They also listen and respond to what their classmates have posted.

Research links

Choe, Y. (2011) 'Effects of task complexity and English proficiency on EFL learner's task production in SMSC', *Multimedia-Assisted Language Learning*, 14, pp. 3–34.

Yilmaz, Y. and Granena, G. (2010) 'The effects of task type in synchronous computer-mediated communication', *ReCALL*, 22, pp. 20–38.

Sample Screenshot

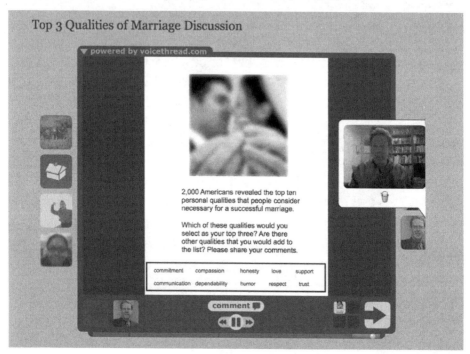

Screenshot courtesy of www.voicethread.com

Activity: Listening Games

Introduction

There are increasing numbers of online language games and game apps that students can play to improve their listening. Many games, which can be replayed again and again, are free or cost very little. Most games also have a degree of flexibility that means they can be played for just a few minutes, whenever a student has some free time. This activity provides some guidance for a selection of games and ways of playing them mindfully, to improve comprehension and retention.

Aim
- improve perception and memory skills
- have fun while listening
- develop these active listening strategies: comprehension monitoring, performance evaluation, multimodal inferencing, repetition

Level Beginner +

Time 20 minutes

Materials

One or more online games

Listening Master: This game requires students to listen to a short illustrated extract and answer global comprehension questions.

http://www.lingual.net/game/listening-master, also available at iTunes app store, search Listening Master

English Wizz: This game requires students to listen and select the best completion to sentences.

http://www.lingual.net/game/english-wizz/, also available at iTunes app store, search English Wizz

Elllo Listening Games: These games require students to follow a story and answer comprehension questions to advance.

http://www.elllo.org/games/student_games.htm

If you want to create your own games from scratch:

Super Teacher Tools: you can use existing games in Jeopardy or Millionaire-type formats or create your own game using the editing tools provided on this site. These games do not use audio.

http://www.superteachertools.com/

http://www.superteachertools.com/boardgame/makegamehtml.htm

Create a PowerPoint of a Jeopardy-style game: with the questions and answers hyperlinked.

How to create a Jeopardy Powerpoint

http://www.youtube.com/watch?v=AR5Zci8vuGY

A lower tech version is to use flashcards, with the category/points on front and the question on the back.

Preparation

- Preview an existing online listening game, from the resources in the Materials section, or another source you know. Play one round of the game yourself, to get a sense of the 'player experience': what decisions does the player make, what physical actions are required, what time pressure is the player under, what thinking processes are involved, what suspense or fun elements are in the game?

- Prepare an online connection to play a round of the game in class, with a monitor large enough for students to view.

Procedure

1. (before listening) Ask the students if they have examples of free (or low cost) listening games that they play as mobile apps or as online games.

2. Navigate to the site of the game on a classroom monitor. Be sure you have adequate volume for the audio.

3. (listening) (For the Listening Master game) Divide the students into two groups, and get them to start the game. Spin the wheel and monitor whose turn is next. Players may consult each other before giving you their final answer.

4. Keep score. Allot points for correct answers and deduct points for incorrect answers. Have students use a simple score sheet to help follow the game (see worksheet).

5. (after listening) At the end of one or two rounds, 'debrief' the students: Was the game interesting? What did they learn from playing? Would they like to play again on their own?

6. As a follow up, have students note the URL of one of the games they would like to play on their own, or discuss how to locate it in an online store. Encourage them to play 20 minutes per week, to improve their listening on their own.

Variation

Classroom games. If you have an online connection in the classroom, you can play online games with your class, perhaps just for a few minutes at the end of class. Divide the students into teams, or have each student record their own answers first before you poll the class for the correct answer.

Comments

In our own teaching, even with advanced students and groups of serious businesspeople or diplomats, we find it useful to have occasional game sessions in a class. Using game-like activities can be a beneficial break from other forms of study, and can serve as valuable learning time as well.

Research links

http://www.teflgames.com/why.html.

http://www.earlychildhoodnews.com/earlychildhood/article.

Crookall, D. (2010) 'Serious games, debriefing, and simulation/gaming as a discipline', *Simulation Gaming*, 41, pp. 898–920.

Dudeney, G., Hockly, N., Pegrum, M. (2013) *Research and Resources in Language Teaching: Digital Literacies*. Harlow: Pearson Education.

Rogers, Y. and Price, S. (2009) 'How mobile technologies are changing the way children learn' in Durin, A. (ed.) *Mobile Technology for Children: Designing for Interaction and Learning,* pp. 3–22. Burlington, MA: Morgan Kaufman.

Zheng, D., Young, M.F., Wagner, M. and Brewer, B. (2009) 'Negotiation for action: English language learning in game-based virtual worlds', *Modern Language Journal*, 93, pp. 489–511.

Worksheet and Sample Screenshot

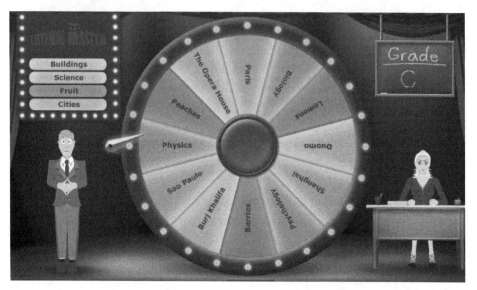

Screenshot courtesy of www.lingual.net/game/listening-master

Category	Question (key words only)	Answer (key words only)	My score (+ / −)

Activity: News Hound

Introduction

Listening to the news in English is one of the more challenging tasks for language learners, and it can be a real measure of progress in understanding authentic language. The news is also something that students – particularly adults – can do on their own once they've had some guidance on how to use certain strategies, e.g. creating a simple task, listening for key words, listening for topic sentences, etc. As with other autonomous listening activities, if the teacher can provide mentoring and feedback, the results are likely to improve.

Aim
- find accessible sources for listening to the news in English
- practise following current events in English
- summarise news stories
- develop these active listening strategies: resourcing, world elaboration, noting, summarisation

Level Intermediate +

Time 25 minutes

Materials

News sources:

http://www.voanews.com/english/news/ or
http://www.voanews.com/learningenglish/home/

http://www.cnn.com/

http://www.bbc.co.uk/news/

http://www.bbc.co.uk/worldservice/learningenglish/

http://www.wherever.tv/tv-channels/IBA-English-News.jsf

http://www.ndtv.com/video/live/channel/ndtv24x7 (India)

http://www.ustream.tv/channel/nhk-world-tv (Japan)

http://www.breakingnewsenglish.com/links.html (simplified news stories)

Preparation

1. Listen to a news broadcast and take notes on the main stories (see the following worksheet). This acts as an answer key. Alternatively, record the TV news. This will provide useful visual support.

2. Listen for difficult essential vocabulary that you may need to pre-teach.

3. If you have access to a language lab, set up each console so that the students can listen alone and independently to the news.

Procedure

1. (before listening) Ask the students what the main news stories of the day are, locally, nationally and globally.

2. Pre-teach any of the words you identified during the preparation stage.

3. (listening) Explain that the students will hear the news and complete the worksheet (at the end of this activity). Play the recording two or three times.

4. Give the students a few minutes in groups to compare their answers.

5. Listen/watch again, this time dividing the broadcast into shorter chunks and providing the answers. Go over any difficult vocabulary as it comes up. (Option) The students can be asked to do secondary tasks: note down any relevant vocabulary they wish to ask about, fill gaps in a transcript, or correct factual errors in summaries of the stories.

6. (after listening) As a follow up, have students identify four news stories to listen to outside of class. They summarise the stories or create a podcast mimicking a news reporter for one of the stories.

Variation

- Intermediate students may benefit from listening to or reading the news in their own language first. If they already know the story, this means there is the mental processing space available for them to focus on the English language. You may need to tell them to do this before class or give them several minutes to do this in class before they listen in English.

- Ask students to work in groups to list the biggest news stories of the previous year. Then get them independently to watch a compilation of the year's news stories on YouTube and to note down which stories came up.

Comments

We have found that one of the trickiest things for students about listening to the news is the speed at which the newsreaders speak. However, if students do this activity repeatedly, they learn to adjust their expectations so that they can keep pace with the newsreader. In other words, they realise that they need to be 'on high alert' once the recording starts. It is easier to watch the news than to listen only; the structured, predictable nature of news broadcasts and the listener/watcher's knowledge of the theme are both strong predictors of listening success. For these reasons, TV sources are also recommended.

Research links

http://iteslj.org/Techniques/Mackenzie-CNN.html.

http://www.cc.kyoto-su.ac.jp/information/tesl-ej/ej27/a2.html.

Donaldson, R. and Haggstrom, M. (2006) 'Wired for sound: teaching listening via computers and the world wide web', in Donaldson, R. and Haggstrom, M. (eds) *Changing Language Education Through CALL*. London: Taylor and Francis.

Worksheet

News source: _____ Date: _____

	What?	Where?	When?	Who?
Story 1				
Story 2				
Story 3				
Story 4				

Activity: Vox Pops

Introduction

One aspect of autonomous learning is finding ways to practise listening and interaction skills outside the comfort of the classroom. In this activity, the students' task is to conduct interviews in English with native speakers or other fluent speakers. Autonomy is developed in planning, preparing questions and gathering the courage to find and initiate interactions with interviewees. During the interview, the students have the opportunity to use the collaboration strategies they have developed (seeking clarification, backchannelling) in a real-life, real-time interview situation. This aspect of connecting with speakers in the community, in a structured and supported way, can be a very productive form of learning for students.

Aim	• listen and converse in real time
	• use strategies for conversational repair, clarification, confirmation, etc.
	• transfer classroom practice to the community
	• develop these active listening strategies: advance organising, comprehension monitoring, emotional monitoring, seeking clarification, noting
Level	Intermediate +
Time	60 minutes

Materials

Possible sources:

http://www.youtube.com/user/PearsonELTSpeakout/videos

http://www.youtube.com/user/LukesEnglishPodcast?blend=1&ob=video-mustangbase

Search YouTube for 'man on the street interviews'.

Preparation

1. Prepare to show some short vox-pop-type interviews (suggestions above) with simple gist questions, for example:

   What do you think of . . . ?

   What's your favourite . . . ?

   How do you spend your . . . ?

   Did you watch (event on TV)? If so, what did you think of it?

2. Decide on the topics the students will interview people about. Alternatively, you can let the students decide.

Procedure

1. (before listening) Show the students some vox-pop-type interviews. Do some comprehension work, and get them to discuss the questions that the interviewer asked.

2. Explain that the students will be conducting interviews in English either with people in the community, if available, or teachers/high-level students in the school.

3. Put the students into pairs and give them the topic. Have the students come up with six or seven topic-based questions for their interviewees.

4. Ask them to check their questions with another pair and consult you if there are any queries about correctness. Then ask the students to narrow down their questions to the three or four best.

5. (listening) Set the conditions for the interview. Some students may be able to film their interviews on flipcams or cell phones, or audio record, or perhaps take a headshot photo of their interviewees. Set a goal, e.g. interview three separate people, and a deadline, e.g. one week, for the interviews to be completed. Also prepare students for how to begin speaking to a stranger (see the worksheet).

6. (after listening) Get the students to report back on their interviews. This stage could include the worksheet to give the feedback stage a clear structure.

Variation

Translation interviews. In EFL environments, where it may be impossible to find speakers of English to interview, the students can conduct their vox pop interviews in the native language of the country. They need to record and translate the upshot of each interview into English and present their results in class.

Comments

In some contexts, finding native speakers or very competent speakers will be challenging. If necessary, consider bringing in expatriates or competent speakers known to you. Also consider setting up telephone interviews with such people. The students are usually pleased to be able to practise their English with members of the native speaker community.

Research links

Russell, N. (2007) 'Teaching more than English: connecting ESL students to their community through service learning' in *Phi Delta Kappan*, 88, pp. 770–771. Bloomington, IN: Phi Delta Kappa International.

Worksheet

Talking to a stranger: a few suggestions on etiquette

1. Stop them politely. ('Excuse me, do you have a moment?')

2. Tell them who you are and where you are from. ('My name is X ... and I'm from ... ')

3. Say what you're trying to do. ('I'm doing a survey on ... ')

4. Ask for their help and apologise for the inconvenience. ('Would you be able to answer a couple of questions?' 'Do you mind if I take a photo/film your response?')

5. Thank them at the end. ('Thank you very much for your help.')

Interview: Feedback Worksheet

What questions did you ask?

What went well during the interviews?

What was difficult about the interviews for you as a listener/speaker?

What would you change about the interview if you could do it again?

What information did you learn?

What new language did you learn?

Activity: Webquest

Introduction

Based on one overarching question, webquests should be an invitation to think. The best webquests require higher order processing – synthesising, analysing and problem-solving – rather than just finding facts. In their fullest form, webquests also require multimodal processing and the use of all four skills: reading, listening, writing and speaking. Because most websites are not produced with language learners in mind, the language will be challenging for students. The literacy goal is for students to navigate their way around problematic authentic material, using cognitive strategies, such as guessing from context, in order to complete the task. Depending on the topic of the webquest, sites with large free archives (e.g. the BBC, YouTube) are particularly useful as resources.

Aim
- work autonomously to achieve a listening task online
- use English internet sources to expand knowledge
- develop these active listening strategies: directed attention, double-check monitoring, resourcing, noting, collaborating

Level Intermediate +

Time 20 minutes

Materials

Websites plus written tasks/questions in a worksheet.

A good place to start building a webquest is http://www.zunal.com/. This will take you step by step through the process.

An example of an excellent webquest is http://www.esllistening.org/beatles/beatles/webquest.html

Preparation

Find a number of websites with a common theme. Prepare a worksheet that requires students to visit the sites in sequence.

Procedure

1. Choose a topic that appeals to your students. Design a series of exercises based on web materials concerning the topic. These exercises must form a coherent sequence, each building on the previous one. Make sure there is a listening component as well as reading. (See the sample webquest in the worksheet.)

2. Distil the quest to one major question, e.g. Who was Che Guevara? Why did The Beatles become the biggest band in the world? How do you make a movie? Some other fairly generic titles can be adapted for different classes: 'A Day in the Life of . . .', 'The Secret History of . . .', 'How to . . .', 'XXX Uncovered'.

3. Write the question or title on the board and do some in-class work to introduce the topic, such as a reading, a discussion or a quiz. Tell the students they will be doing the task online independently. Assign a deadline. If you have put the webquest online, direct the students to the site. Remind them that multiple listenings are necessary to assure comprehension.

4. Include a feedback stage that goes over the answers and incorporates discussion about what else the students learned from the activity.

5. As a follow up, the students do a group presentation, to amplify the listening outcome.

Comments

We have found that during the feedback stage it is useful to mention various metacognitive and cognitive strategies. We ask students if it was necessary to understand everything in order to complete the task. We also ask them how they dealt with difficulties: did they use dictionaries, listen again to recordings, or look at information from other sources in order to confirm their hypotheses?

Research links

http://webquest.org/.
http://webquest.sdsu.edu/about_webquests.html.
http://zunal.com/.

Worksheet

Sample Webquest

Question: Who was Che Guevara?

Questions	Links
What are the bare facts of his life? Dates: Places: People in his life: Achievements: Legacy:	http://www.biography.com/people/che-guevara-9322774
What is the myth?	http://www.youtube.com/watch?v=fqTw2dtVQzw
Why is he a controversial figure? Arguments for him:	http://www.youtube.com/watch?v=-Q9KgeCDdOo
Why is he a controversial figure? Arguments against him:	http://www.youtube.com/watch?v=JNZ5MnKDLnE

Activity: My Listening Library

Introduction

This activity comes under the umbrella of **extensive listening**, which we consider listening for extended stretches of time (three minutes or more) without stopping. Though this length of listening may seem too challenging for many learners, it's important to realise its value. If a student can practise extensive listening, in bursts of a few minutes, for even half an hour a day, this practice is sure to yield results in confidence, increase in attention span, and overall proficiency gains. The key is in using strategies that help students to plan, to regulate their attention and to deal with comprehension problems while listening.

Aim
- find sources for extended listening outside the classroom
- practise fluency listening
- develop these active listening strategies: self-management, persistent attention, retrospective inferencing

Level Intermediate +

Time 30 minutes (the time to set up the out-of-class activity)

Materials

Websites that provide free online listening resources. Keep a list of your recommended resources on a course website or in a readily accessible folder.

Regularly updated sites that contain reputable listening material: news broadcasts, documentaries, interviews, dramatic stories. Possible sources include:

Intermediate

Spotlight Radio: http://www.spotlightradio.net uses a specially modified form of English to make listening to the radio easier. You can listen and read at the same time.

Voice of America – Special English:
http://www.voanews.com/specialenglish/index.cfm These are broadcasts in simple English (only 1,500 different words) on hundreds of topics.

Student News (CNN): Weekly news summaries, international and national (US) news
http://podcasts.cnn.net/cnn/big/podcasts/studentnews/video/

ELLLO – English Language Listening Lab Online http://www.Elllo.org. There are over 1,000 free online listening activities for teachers and students. Many activities have quizzes and a transcript.

Short stories with selected sentences for dictation practice: http://www.eslfast.com/#A

Selected YouTube videos, subtitled by members http://yappr.com.

Connect with English: http://www.learner.org/resources/series71.html A video instructional series in English as a second language for college and high school classrooms and adult learners; 50 15-minute video programs and coordinated books

Advanced

Broadcasts from the BBC – a range of audio and video broadcasts available for online streaming: www.bbc.co/uk/worldservice/BBC_English/progs.htm

Free TV – watch your favourite TV shows. Any time. For free. http://www.hulu.com/

Awesome Stories – the story place of the web; tell yours, listen to others. www.awesomestories.com

Soundprint Extended feature stories. Sample: http://soundprint.org/radio/display_show/ID/823/name/Time+on+the+Outside%3A+Hope's+Story

History Channel Speeches http://historychannel.com/speeches/index.html – many famous historical speeches.

BBC World Service – Learning English: http://www.bbc.co.uk/worldservice/learningenglish/multimedia/. This site has hundreds of listening activities and texts from the BBC's extensive archives.

World News Australia: www.youtube.com/user/worldnewsaustralia.

Preparation

Collect a small archive of online resources from which your students can select what they will place in their own 'library'. To save time, it is best to start small and build up 'organically', using the students' own recommendations. Create a list of five sources of listening material that students can access. This may include graded readers with audio CDs or downloadable mp3 files that are purchased for the class to use on a circulating basis.

Procedure

1. (before listening) Explain to your students what extensive listening is: listening *comfortably* in English for several minutes *without stopping* the audio or video. Explain to the students the benefits of extended listening outside of the classroom: it will help them become comfortable listening (even when they don't understand everything) and confident about their ability to get the main ideas. Extended listening is an effective way to 'think in English'.

2. Write these guideline questions on the board or include in a handout or on the class website:

   - Can I understand *about 80%* of the content?
   - Can I understand *most* of the vocabulary and grammar?
   - Can I listen and understand *without having to stop* the audio or video player?
   - . Am I *enjoying* the content of the listening material?

   Coach the students to create their own guidelines:

   - If your answer to all these questions is *yes*, then you have found the right level for yourself!
   - If your answer to any of these questions is *no*, then it might be difficult for you. (Challenge is good, but, for extended listening, if you get frustrated or tired, you won't enjoy the activity and you may stop listening.) Also, if you don't enjoy the content of the listening material, you'll soon become bored, so choose something interesting.

3. Present your list of resources that the students can choose from. Start with just a few. Ask the students if they have other ideas to contribute. If you have online access in the classroom, go to the actual online sites and demonstrate the resources.

4. Ask the students to form teams who will exchange the same resources.

5. (listening) Distribute the listening logs (see the worksheet). Explain how to use them: at the end of a session (aim for 30-minute sessions, at least one a week), fill in the date/time, the source, etc.

6. (after listening) Spend some class time regularly to check on students' progress and to share tips for extended listening. Remove resources that none of the students likes and add resources based on student recommendations. Remember: the goal is to find *engaging* listening materials that the students *enjoy* listening to!

7. (Optional) Once every week or two weeks, have your students bring their listening log to class, to share their top resources with the class. Spend 10–15 minutes sharing hints and strategies for extensive listening. At this time, you can add any new resources you've found. This type of updating can also be done through a course website or Facebook page.

Variation

Repeated listening. An alternative to extended listening or listening to many extracts one time only is repeated listening, which refers to listening to the same extract multiple times, with repeated tasks, such as note-taking, or varying tasks, such as answering different types of questions. To do this, create two or more tasks for each extract, gradually increasing the difficulty and level of detail.

Comments

We have found it is best to start small with this type of activity, and allow it to grow. You'll find that many students begin to make remarkable progress when they engage in extensive listening. Even after your course is over, if the students have tasted some success with extensive listening, they are likely to continue to gain in listening proficiency.

Research links

Rendaya, W. and Farrell, T. (2011) 'Teacher, the tape is too fast! Extensive listening in ELT', *ELT Journal*, 65, pp. 52–59.

Woodall, B. (2010) 'Simultaneous listening and reading in ESL: helping second language learners read (and enjoy reading) more efficiently', *TESOL Journal*, 1, pp. 186–205.

Rob Waring's website on extensive reading and extensive listening: http://www.robwaring.org/el/.

Worksheet

Extensive Listening Log

Fill out the calendar for the current month. Make symbols on each day of the calendar:

1. The amount of time you listened today:

    < = less than 30 minutes   >> = more than 60 minutes

    > = 30–60 minutes

2. What did you listen to?

    N = news   M = music

    D = documentary   S = story

    F = film   I = interview

3. How was your listening experience today?

    ++ very good   – bad

    + good   — very bad

    ~ OK

Sun	Mon	Tues	Wed	Thurs	Fri	Sat

For at least one listening experience each week, fill out this chart

What did you listen to? Give a general summary	What was most interesting to you?

What did you learn?	Recommend? 😊😊 Any other comments?

Activity: Learn Something New

Introduction

In order to make progress in listening, students have to challenge themselves. They can do this by attempting to listen to more difficult passages,
by listening multiple times to notice more, or by listening narrowly – to multiple extracts on the same topic to develop deeper knowledge. Students are
likely to be motivated to deal with challenging input in their L2 if they have
a strong desire to learn something new. To build up and sustain the students'
desire, this activity involves four phases: whole class work (for surveying the
choices), individual work (for watching or listening to selected videos), small
group work (for discussing and summarising), and large group work (for sharing their report).

Aim
- challenge yourself with new content

- learn something new in English

- develop these active listening strategies: planning, collaborating,
  reviewing

Level Intermediate +

Time 20 minutes to introduce the activity

Materials

Select a number of online listening resources of varying lengths. These can
be on topics of general or specific interest, or current news topics. The key
to selecting extracts is to provide access to new knowledge or perspectives.
Cut to reduce to 8–10.

Here is a list of materials to try:

Intermediate level

http://www.exploratorium.edu/listen/lg_michael_video.php
(Michael Chorost: Electronic Listener)

http://www.exploratorium.edu/listen/lg_bart_video_sub.php
(Bart Hopkin: Experimental Musical Instruments)

http://www.exploratorium.edu/listen/lg_doniga.php (three minutes)
(Doniga Markegard: Wildlife Tracker)

http://www.khanacademy.org/
(Khan Academy Learn almost anything (mostly about science))

suggested:

http://www.khanacademy.org/video/introduction-to-evolution-and-natural-selection?playlist=Biology (Introduction to Evolution and Natural Selection – contains optional closed captioning)

http://academicearth.org
(Academic Earth Online courses from the world's top scholars)

http://www.youtube.com/watch?v=gLBE5QAYXp8
(The Story of Stuff: entertaining and accessible talk on the environment)

Advanced level

http://www.khanacademy.org/video/ponzi-schemes?playlist=Finance
(Ponzi Schemes)

www.academicearth.org

http://www.learning2010.com/Videos/apolo-ohno.htm
(Short Talk by Olympian Apolo Ono about his philosophy of Zero Regret)

http://www.ted.com/talks/alexander_tsiaras_conception_to_birth_visualized.html (Alexander Tsiaras: Conception to birth – visualised)

http://www.ted.com/talks/lang/es/isabel_allende_tells_tales_of_passion.html (Isabel Allende: Tales of Passion – start at 7:28 to 11:16, or 11:17 to 13:52)

Preparation

Skim through the videos before you put them on the list for the students. Create a short list of videos in categories of difficulty. You can always add to the list later.

Procedure

1. (before listening) Select 5–10 listening sources from the Materials list above or include your own sources. Provide a brief description of each one you decide to include.

2. Distribute the lists, or use a google document <document.google.com> for students to share.

3. Provide a short summary of each talk or speculate briefly on what you think it's about. If you have an online connection, you can show the first 15–30 seconds of each talk to provide a 'taste'. Note that some videos have closed captions (cc).

4. First have students rate each topic in terms of interest:

   3 = very interested

   2 = somewhat interested

   1 = only a little bit interested

   0 = not interested at all

5. Get the students to circulate and find other students who have 3s for any topic. They can ask, 'What are your most interesting topics?'

6. (listening) In pairs or small groups, the students plan to watch the selected clip before the next class, take notes and come to class ready to discuss the main ideas with their group.

7. (after listening) As a follow up, have the students meet in their small groups to prepare short presentations of the clips. Use the 10-word summary provided in the worksheet to keep talks focused. Allow time for questions at the end.

8. (Optional follow up) Get students to watch one another's videos and then discuss if they agree or disagree with the ratings.

Variation

Interactive transcript. Where available, show students how to use the **interactive transcript:** they can click on parts of the transcript to jump to a particular part of the talk to hear those parts again.

Comments

We have found, both as teachers and as language students ourselves, that learning content in an L2 is very rewarding. The feeling of authenticity ('I'm actually connecting with this speaker or writer in their native language!') and eventual triumph trumps the effort that is required. If you can help students engender this feeling, they will want to keep expanding their autonomous learning in their L2. This idea is the basis of CLIL (content and language integrated learning), which became popular in Europe in the 1990s as a form of language immersion that could help schoolchildren, in particular, acquire English while learning school subject matter.

Research links

http://clilrb.ucoz.ru/_ld/0/29_CLILPlanningToo.pdf
(Planning tools for teachers).

Chang, A. and Read, J. (2007) 'Support for foreign language listeners: its effectiveness and limitation', *RELC Journal*, 38, pp. 375–394.

Chung, J. (2002) 'The effects of using two advance organizers with video texts for the teaching of listening in English', *Foreign Language Annals*, 35(2), pp. 231–241.

Elkhafaifi, H. (2005) 'The effect of prelistening activities on listening comprehension in Arabic learners', *Foreign Language Annals*, 38, pp. 505–513.

Herron, C., Hanley, J. and Cole, S. (1995) 'A comparison study of two advance organizers for introducing beginning foreign language students to video', *Modern Language Journal*, 79, pp. 387–395.

Iimura, H. (2007) 'The listening process: effects of types and repetition', *Language Education and Technology*, 44, pp. 75–85.

Sakai, H. (2009) 'Effect of repetition of exposure and proficiency level in L2 listening tests', *TESOL Quarterly*, 43, pp. 360–372.

Worksheet

Rating Form

Source	Ten-word summary	My rating	If a '3', who has the same rating?
Michael Chorost: Electronic Listener	Deaf man describes how cochlear implant helps him hear.		
Apolo Ono: Zero Regret			
Alexander Tsiaras: Conception to birth – visualised			
Isabel Allende: How I write a book (cc)			
Khan Academy: Ponzi Schemes			
Khan Academy: Introduction to Evolution			
Bart Hopkin: Musical Instruments (cc)			
Doniga Markegard: Wildlife Tracker (CC)			
Isabel Allende: Tales of Passion			
The Story of Stuff			

Report Form

We listened to

It's about

The main ideas of the talk are

One very interesting idea in this talk is

We recommend/don't recommend it because

Activity: Film Review

Introduction

Students naturally will become more active learners outside the classroom if they can find ways to connect language learning to their current interests. Most students have an interest in films of one sort or another: action films, adventure, animation, documentaries, shorts, comedies, historical, musical, science fiction, and so on. There is almost always one type of film that each student *loves* and will be happy to incorporate as a resource for listening development. However, most students will need some support and mentoring in order to maximise the learning opportunity. This activity provides some structure to enable students to get as much language learning as possible out of the films they watch.

Aim
- find sources for listening outside the classroom
- develop strategies for viewing long films in English
- practice fluency listening
- develop these active listening strategies: multimodal inferencing, contextual inferencing, resourcing, mediating

Level Intermediate +

Time 30 minutes (the time to set up the out-of-class activity)

Materials

Starting list of films (these films have been recommended by language students):

2001: A Space Odyssey	*Ratatouille*
A Beautiful Mind	*Thank You for Smoking*
Billy Elliot	*The King's Speech*
Black Swan	*The Piano*
Finding Nemo	*Titanic*
Mary Poppins	*Toy Story*

Preparation

Collect a small archive of DVDs and online resources that contain films you believe are appropriate for your students, in terms of content and language density. Often films with less dense language are more appropriate. You can read the scripts of many films online (e.g. www.script-o-rama.com). Make a list of these films for your students, starting with perhaps 10–15 films, listing the title and a short synopsis.

Procedure

1. (before listening) To familiarise the class with the concept of film reviews, select three current films in English. Ask the students for suggestions. (Try to get films in different genres.) As a preparatory homework assignment, have students collate reviews of these three films, using at least two established film review sites:

   - Rotten Tomatoes http://www.rottentomatoes.com/ (100 point scale)
   - Film Critic http://www.filmcritic.com/ (five star maximum)
   - Screen Jabber http://www.screenjabber.com/ (five star maximum)
   - Mr Cranky http://www.mrcranky.com/ (five bombs minimum)

   Tell the students they need to visit two sites that have reviews of these films and complete the review table (see the worksheet for these films). (Optional: students can visit sites using their L1; have them work at rendering these reviews into colloquial English.)

2. (listening) To familiarise students with the role of trailers for previewing films, have them visit at least two sites that feature free film trailers:

   - Trailer Addict http://www.traileraddict.com/
   - iTunes Movie Trailers http://trailers.apple.com/trailers/
   - DejaVuRL http://www.dejavurl.com/
   - Movie Trailers http://www.moviestrailer.org/

   Tell the students to find and view trailers for the three movies the class is reviewing this week.

3. (after listening) Have the students bring their notes to the next class and complete the review form (see the worksheet).

4. Select the film that has the best reviews. Assign this film for viewing for the following week. (Option: have the students select the film they would like to review; they can form film review groups for the same films.)

Variation

Subtitles on/off. Encourage students to watch the film once with English subtitles and then again without subtitles. Most DVD menus are indexed by scenes and most controllers will allow you to turn subtitles on and off easily, so you do this method scene by scene. Watch once with subtitles, then skip back to the beginning of the scene and watch it without subtitles.

Comments

We have found that films, like music, can offer a wonderful enhancement to language learning. While virtually any film can afford material for language development, we have found it's best to steer students away from films that have high verbal density (lots of speaking and very little action), high action density (lots of action and very little speaking), high sensationalism factor (too much dependence on violence and special effects), and low language learning relevance (lots of technical language, period language, regional language). And interestingly, when students are given the opportunity to rate films in terms of value, they generally will identify these 'utility factors' on their own.

Research links

Clark, R., Nguyen, F. and Swellen, J. (2006) *Efficiency in Learning: Evidence-based Guidelines to Manage Cognitive Load.* San Francisco: Wiley.

Jones, L. and Plass, J. (2002) 'Supporting listening comprehension and vocabulary acquisition with multimedia annotations', *Modern Language Journal*, 86, pp. 546–61.

http://www.video.about.com/

http://www.yappr.com

http://www.jokeroo.com/shorts

http://www.futurethought.tv

Worksheet

Film Review Form

Name of film

Short summary of the film

What rating did this film receive on:

Rotten Tomatoes	Film Jabber
Film Critic	Mr Cranky

How would you rate this film? (1 to 5 stars)

Interest

Story

Good for learning language

Activity: Conversation Corner

Introduction

In order to assist students in becoming active listeners and active learners, we have to help them find ways to practise listening and interaction in the L2 outside the classroom. While sources of one-way listening are becoming increasingly easy to access, sources of two-way interactive listening are less easy to identify. One of the most challenging obstacles L2 learners face is finding and sustaining mutually satisfying relationships with conversational partners. This activity outlines ways of setting up chat centres and taking steps to ensure they are rewarding for students.

Aim	• find sources for interaction and listening outside the classroom
	• practise conversational listening
	• develop listening and interaction strategies
	• develop these active listening strategies: collaborating, personal elaboration, emotional monitoring, speaker inferencing
Level	Intermediate +
Time	20 minutes (the time to prepare the activity in class)

Materials

Conversation topic cards (see worksheet)

Openers and topic development phrases

- Hi, I'm . . . Would you like to talk for a few minutes?
- OK, here's a topic . . . What do you think about . . . ?
- Let's talk about . . .
- What does . . . mean to you?
- Do you have any experience with . . . ?
- What's your current view on . . . ?

Feedback phrases

- I see what you mean.
- That's an interesting idea.
- That's cool!
- Hmm . . . I've never heard that before.

Clarification and expansion phrases

- Can you tell me more about . . . ?
- I'm not sure exactly what you mean. Can you explain that again?

Preparation

1. You will need a location where you can hold your chat corner. Find a public area near the classroom facility that is accessible to most students. This may be a section of the student lounge or a room that is rarely used. You may also identify a nearby park or public square.

2. Prepare some topic cards that contain general topics that can be developed for 5–10 minute conversations (see worksheet). The idea is to make the conversation something of a game. The participants need to generate enough questions and sub-topics to keep a conversation going for at least five minutes before going on to a new topic or finding a new partner.

Procedure

1. (before listening) Talk to your students about the value of getting listening, interaction and conversation practice outside of class. Survey the students casually about what kinds of extra practice (a) they are now getting and (b) may be beneficial to them, but are hard to find. Nearly always, the idea comes up of a conversation centre or English circle where students can practise spoken English.

2. Elicit ideas for creating a conversation corner that students can access in their free time. Where is a good location? What is needed to make it work? (Refreshments? Board games? Computers? Volunteer teachers? Volunteer native speakers? Volunteer non-native speakers of English?) What kind of organisation is needed? (Monitors? Teachers? Rules? Schedules?) When can we start? How can we assess its success? How can we make adjustments when needed?

3. Introduce the idea of conversation topic cards as a way of keeping conversations fresh, and avoiding repetition of the standard getting-to-know-you conversations (which can become tiresome and lead to student attrition).

4. (listening) Practise the conversation topic cards idea in class. Make cards and have students rotate with different partners, selecting a conversation card and making a conversation about that topic for three to five minutes. When you say 'change', ask students to change partners and choose a new card.

5. (after listening) After four or five turns, debrief the students. What did they think of the activity? How would they adjust it to make it better? Would this kind of idea work in a conversation centre with people you don't know? How could this type of procedure be adapted to a conversation centre?

Variation

Online conversation partners. There are numerous services that will pair an English language learner with a conversation partner, a trained native English speaker, personal tutor, a volunteer, or a language exchange buddy for online conversation practice via Skype or a webcam-based exchange. Usually these will entail a fee. It is possible, with a little research, to locate a school or college programme that may be interested in an online language exchange programme.

Comments

When we visit language programmes in different countries, we often see creative ideas for starting and maintaining conversation corners and various kinds of extracurricular meetings for allowing students to use English conversationally outside the classroom. Examples run the gamut from basic benches with soft drinks to fully fledged entertainment lounges with a stocked (non-alcoholic) bar! Most of the successful corners started small and allowed the idea to take on a life of its own. You'll find that, if a few students take the lead, the chat corner is likely to expand at a suitable pace. Even after your course is over, if the students have experienced success with the conversation corner, they are likely to continue to participate, or find ways to extend the concept into their own development in some way.

Research links

Bardovi-Harlig, K. (2010) 'Recognition of conventional expressions in L2 pragmatics' in Kasper, G., Nguyen, H. and Yoshimi, D. (eds) *Pragmatics and Language Learning*, Volume 12. Honolulu: NFLRC.

Weinart, R. (1995) 'The role of formulaic language in second language acquisition: a review', *Applied Linguistics*, 16, pp. 180–205.

Worksheet

Conversation topic

Relationships	*Current issues*	*Talents and skills*	*Friends and relatives*
Interests and recreation	*Childhood*	*Health*	*Work and career*
Adventure	*Movies*	*Ambitions*	*Personality*
Technology	*Holidays*	*Games and sports*	*Travelling*

Worksheet

Conversation Corner (CC) Schedule

When is the corner open? Who will be there at each time?

Sun	Mon	Tues	Wed	Thurs	Fri	Sat

Every time you visit the CC, fill out this log

Date and time
Who did you talk to?
What topics did you talk about?
What expressions did you learn or what new expressions did you use?
Any suggestions to improve the CC?

Part III

From Application to Implementation

This section describes how to integrate the activities in Part II into an overall curriculum. It looks at ways to adapt the activities to fit numerous teaching situations and contexts, and then describes how teachers can generate their own activities based on the underlying principles of the five frames (Affective, Top Down, Bottom Up, Interactive and Autonomous).

Integrating active listening into an overall curriculum

Numerous questions arise when we consider integrating active listening into a curriculum. We have organised these questions under six broad headings:

1. Choosing the content of the listening curriculum

2. Organising the listening curriculum

3. Integrating listening with other skills

4. Adapting the listening curriculum: institutional constraints and opportunities

5. Adapting the listening curriculum: different student populations

6. Adapting and creating the listening curriculum: different teachers

Where the questions deal with practical classroom issues, we refer back to the activities in Part II.

Choosing the content of the listening curriculum

What is the role of listening in the language curriculum?

In any language learning curriculum, listening has multiple roles to play. One of these roles is simply to practise the skill of listening – effectively, listening for listening's sake – or what we might call pure skills development. This approach involves using recordings of factual or fictional passages, conversations, video clips, teacher monologues, etc. The content needs to be of high interest, and the pedagogical focus is on learning to cope with the features of spoken language through the use of strategies, such as listening selectively and anticipating content.

With skills development lessons, any new vocabulary and grammar is acquired incidentally. In other words, acquisition of language takes place as a side effect of the activities rather than being the goal. The true objective of listening skills development is to improve the students' ability holistically – to process more language, delivered at higher speed, involving increasingly challenging grammar, vocabulary, topics, genres and speaking styles. Many language specialists consider this holistic development to be the true measure of second language (L2) acquisition.

How can I help students learn language through listening?

Teachers often use listening in order to help students learn language, specifically new vocabulary and grammar. This means the listening passages contain embedded grammar and vocabulary, which the students are primed to notice. The recordings produced in order for students to learn language in this way are usually scripted. The scripts include repeated or exaggerated examples of the target form presented in a semi-realistic context. In this type of extract, the genre and situation may be natural, but the actual words and delivery may come across as 'acted'.

Here the focus is on new language points rather than the skill of listening. Usually, there is a post-listening activity involving pulling out the target forms rather than dwelling on problems the students may have had while listening or phonological features of the recording that made it challenging to understand. Most veteran teachers will remember course book dialogues that contained dozens of examples of *will* or *used to* or *have to* as exemplars of target grammar. As *listening material*, these conversations sounded highly unnatural, but as *grammar teaching material*, they worked as intended.

What is the role of teacher-talk in listening?

Teacher-talk refers to everything that the teacher says in the target language. This is, of course, an important source of listening practice! (Student-talk refers to everything that the students say in the target language – and this, too, is a vital source of listening input for the class!) Teacher-talk can come in many forms, for example explanations, directives, instructions, anecdotes, monologues and stories. Some teachers have no option but to rely on teacher-talk as the main source of listening input, while others decide not to use any commercially produced recordings or clips, adopting a method that has been called 'teaching unplugged'. Why might teachers take this technology-free option? These teachers see 'authenticity' as the live communication between teacher and student (and student–student) and place little value in bringing inauthentic voices into the classroom. In this view, scripted speech is not the same as real speech, and the students' listening goals usually involve understanding authentic speech.

There are numerous differences between authentic and scripted speech. Authentic speech tends to be faster, contain more repetition and redundancy, and include more non-standard forms (e.g. 'it ain't'; 'I can't find no paper') and colloquialisms. Authentic conversations are usually longer and contain misunderstandings, false starts, and communication breakdowns and repair. Scripted speech, on the other hand, tends to present a 'can-do' world in which transactions of all kinds go smoothly, and communication generally succeeds without problems. Standard forms are used, people speak clearly and seldom have to self-correct, and they move from topic to topic in the conversation without 'wandering off track'! As a consequence, scripted speech tends to be easier for students to understand. Indeed, many are surprised by just how difficult it is to interact with native speakers in the country where the target language is used because authentic speech sounds nothing like the scripted recordings they hear in their language classes!

What are the advantages of teacher-talk?

The use of teacher-talk as a primary source of instruction means teachers no longer have to worry about finding listening material because everything is contained within the minds of the students and teacher. Using only 'live talk' also alleviates concerns over failing technology and power cuts. Another advantage is that the teachers can grade their language according to the students' level and make adjustments in the form of explanatory comments or rephrasings – on the spot – in response to their students' problems with comprehension.

Are there any drawbacks to using teacher-talk exclusively?

Perhaps the biggest drawback of using only teacher-talk and student-talk is the lack of variety and therefore the lack of exposure for the students to a fuller range of the target language. Bringing other voices into the classroom, whether through course book recordings or authentic material, can open the students' minds and ears to the huge variations of global English – for example, Indian English, Australian English, Chicano English, Cockney, pidgin and Creole. In addition, there are other variations: adults, teens and children all have different ways of speaking.

What's more, an outside voice may provide content that is beyond the students' and teacher's capabilities: lectures on history or geography, news broadcasts from around the world, a dramatic film scene. For these reasons, we advocate extensive, but not exclusive, use of teacher-talk and student-talk. Here are some activities from Part II that involve both: *False Anecdote*; *Memories*; *Photo Album*; *Milestones*; *A New Skill* and *Bucket List Bingo*.

Should the teacher speak to the students in the target language all the time?

When you, as the teacher, speak to the students in English all or most of the time, you are using what we call 'informal real-world listening' or 'spontaneous input'. Your casual chit-chat and also your directions and feedback provide an important source of authentic input for the students. Even if you are not a native speaker or proficient speaker of English, your ability to communicate in spoken English is a valuable model for your students. This type of listening is not surrounded by the paraphernalia of classroom listening activities (comprehension questions, answers, formal feedback), but it is every bit as important. Indeed, it can be argued that this is the most authentic form of listening because it involves genuine communicative needs.

Which type of listening should make up the bulk of the curriculum?

A principle of any curriculum focusing on active listening is to provide a balance of different types of listening. The first type – listening skills development – is the focus of the activities in Part II. In addition to this type of listening practice, we acknowledge the value of listening for language learning (many of the activities in the Bottom Up Frame focus on this) as well as the importance of 'live' teacher talk (many of the activities in the Affective Frame feature this) and the immense value of student–student talk (many of the activities in the Interactive Frame feature this type of listening). Variety is essential. Exposure to different genres and different types of text – lectures, casual chat, teacher-talk, TV programmes, film clips, radio shows, recordings, simplified scripted material, and so on – will increase the students' breadth and range of listening ability.

Is it important to teach listening strategies?

We define a listening strategy as *any conscious plan that a learner uses to improve* his or her comprehension or listening performance. Listening strategies that are used by individuals operating in their first language (L1) are not always used in L2. No one really knows why this is the case, but it may be connected to the stress of trying to process a foreign language in real time. In any case, we believe it is advantageous to teach listening strategies, whether directly (i.e. naming and demonstrating the strategy) or indirectly (i.e. coaching our students on ways to improve their listening without naming these ways).

The first task of the teacher is to work out which strategies are teachable. For a strategy to be teachable, it needs to be demonstrable (i.e. a *conscious* plan, such as using visual clues to predict content). It also needs to be repeatable – something that the student will use again and again with different input. For example, students can make informed guesses about unknown words in any context. Finally, a listening strategy must be something that students can incorporate into their behaviour, i.e. the strategy can be adopted and adapted to the students' particular circumstances. The goal, then, is to get students using strategies autonomously and regularly.

Even without studying the eight categories and examples of strategies in Appendix 1, teachers can find ways to help students develop useful strategies. In almost every listening activity, the teacher can give students tips for planning, focusing their attention, monitoring their understanding, evaluating how well they have understood (and what they still need to work on), making informed guesses (inferencing), filling in gaps (elaborating), asking for assistance (collaborating), and reviewing what they have heard. Depending on the needs of the students, the teacher may want to focus on just a couple of strategies that will help them most.

Organising the listening curriculum

How should the listening content be organised?

One option is to provide a 'topic-based' or 'theme-based' curriculum. If we imagine that the theme for the first week is food, the teacher might show a clip of Gordon Ramsay's *Kitchen Nightmares* one day, listen to a food programme on the radio the next day, ask students to listen and mime the instructions for the teacher's home-made recipe for chicken tikka masala the next day, and finish off the week's work by getting the students to conduct a survey on eating habits among the local community. The next week, the class moves on to another theme – perhaps leisure, sport or travel – and again the teacher provides a balanced selection of listening content.

What are the advantages of a topic-based curriculum?

If the topics are well-chosen, the students might find the course stimulating and motivating. This is particularly true if the students are involved in the choice of topics. This can be facilitated by surveying the students at the beginning of a course, thereby finding out what they are interested in and what topics they need to cover for professional or personal reasons.

Are there any disadvantages to a topic-based curriculum?

One potential problem is something known as 'topic-fatigue'. This is when the students get tired of dealing with the same subject, be it in discussions among themselves, recordings or related readings. There are a limited number of approaches to certain topics, and the danger is that the students find themselves rehashing discussions they had a few days earlier on the same topic. One solution is for the teacher to take a broad view of the topic. For example, if the topic is travel, this can include transport, descriptions of places, customs and traditions, intercultural communication, journeys, tourism, travel tips, and so on.

Another potential problem is finding adequate and sufficiently varied materials. This can be an issue for teachers and schools with few resources. Many schools around the world rely on a course book and have little or no access to the internet or TV and radio programmes. While some teachers may be able to supplement course book material by finding their own outside the school, the majority of teachers around the world have neither the time nor the resources to do this. One option, in this scenario, is to use the students as resources, inviting them to supply some of the content of the lesson through anecdotes, demonstrations, articles or recordings that they find, or even English-speaking contacts that may be willing to visit the classroom.

Can a listening curriculum be organised by genre?

Some language programmes have elective genre-based courses that focus on one particular type of listening input: English through film, English through music, English through the news or English through presentations (such as TED talks, which are short presentations given by innovators in many fields and are available on YouTube and at www.ted.com.). Genre can also refer to subject matter. For example, in EAP and ESP courses (English for Academic purposes and English for Special Purposes), the genre may be English for lawyers, English for medical practitioners, English for pilots, etc. A variation on this type of genre-based course is when the students take a course that switches genres on, say, a weekly basis. While not common, this has been done on some listening programmes to good effect. Some students have reported that they enjoy learning through different genres and find it valuable as a way to go beyond their usual language-learning routines. While inevitably not every genre suited every student, there was something there for everyone: songs for the musical learners, film clips for visual learners, and so on.

What are the advantages of a genre-based curriculum?

The main advantage is that the students get intensive practice in listening to the genre. If, for example, they spent a period of time watching news clips every day, there is a good chance that, by the end, they would be able to cope with aspects of the genre because of their growing familiarity with its conventions, its vocabulary and its rhythm. Through repeated exposure to genres, we learn to adapt our listening behaviour and to put appropriate strategies into use, depending on the genre. For example, students who watch documentaries in the target language will soon understand the importance of using the images in conjunction with the aural input to make meaning. They will also pick up on the conventions of the documentary genre, anticipating that the opening few minutes will consist of background information outlining time, place and theme, and perhaps introducing the main protagonists. The more we listen or watch, the more predictable the genre becomes. And predictability is a crucial factor for listeners.

Are there any disadvantages to a genre-based curriculum?

The disadvantage is that the students get no practice in the different types of listening that they will encounter outside the classroom. A course consisting solely of news broadcasts will not help students to practise real-time face-to-face negotiating, for example. While a course that changes genre on a weekly basis *does* provide variety, it does *not* necessarily provide reviewing and recycling stages for each genre. If a three-month course involved, say, listening to jokes only in the first week, by the last week of the course many of the gains in understanding humour in the target language may have been lost because of natural attrition regarding the memory. The curriculum therefore needs to be carefully thought out so that it provides good coverage of many genres. Secondly, it needs built-in opportunities for recycling and revisiting these genres.

Are there any other ways to organise a listening curriculum?

A third organisational option is pure serendipity – using what is available at the time. We might call this an availability-based curriculum. This type of organisation is a daily reality for institutions that have few resources. In various parts of the world, we have seen teachers using whatever listening material they can get their hands on – 30-year-old cassettes, crackly radio broadcasts, passing tourists. These sometimes produce remarkable results because – as has frequently been observed – teachers with poor resources are often the most creative, and students in these situations can be very motivated and engaged.

Can this type of adventitious listening be organised and structured so that it becomes a curriculum? With some ingenuity on the part of the teacher, yes! There are many ways to organise even seemingly random items of listening input. These include: level, based on increasing difficulty and length of the passages; topic or genre; number of speakers – monologues followed by dialogues followed by discussions with multiple participants; and media (radio programmes, course book recordings, face-to-face discussions), to name a few. For a listening curriculum to have coherence, some kind of organisational principle needs to be applied. This principle will allow for the sense of progression and development that is a hallmark of any organised course.

How much listening is 'enough'?

This question depends on several factors. How frequently do the students meet? For what length of time? What other skills and language do they need to cover during the course? What type of equipment is available?

Because of the variability that such questions imply, it is impossible to be prescriptive about the frequency with which students should listen to the target language. However, we can say with some certainty that, in order to improve at the skill, students need to listen *regularly*. Research suggests that three to four hours a week of exposure to input – written and spoken – is necessary to maintain progress in a new language. Only with regular exposure to the spoken language will students get used to the speed at which speakers talk and the accompanying phonological features (elision, assimilation, etc.). Dealing with extended pieces of spoken discourse is, to an extent, a question of practice: it takes time to learn how to concentrate and to develop such strategies as perseverance and using discourse markers to get back on track.

Integrating listening with other skills

Is it better to integrate active listening with other skills or to focus exclusively on listening?

Teachers and curriculum planners often wonder if it's better to have a stand-alone focus on listening or use an **integrated skills approach**. There is no single answer to this, but, in general, an integrated skills approach may be most effective. As illustrated by the activities in Part II, most listening activities involve the use of at least one other skill: speaking, reading or writing. Indeed, frequently a good listening activity is also a good speaking activity, and very often has strong potential with a follow-up reading or writing activity. The question is whether the integration of the skills should be deliberate and balanced, with a percentage of class time dedicated to each skill, or whether it is better to isolate the skill of listening.

To answer this, we need to look at the programmes used by language learning institutions. While some schools teach General English, in which it is the teacher's responsibility to provide a balanced, four-skills curriculum, other schools have separate courses for each skill (as well as separate courses for grammar and vocabulary). Our recommendation is that, whenever the opportunity for a listening portion of a lesson comes up, the teacher aims for an active listening orientation to that portion of the lesson.

What are the advantages of the integrated skills approach?

The clearest advantage of an integrated skills approach to listening is variety within the lesson. Along with many educational theorists, we believe lessons should be divided into segments of approximately 15 minutes, each new segment heralding a change of focus – perhaps a change of skill or a new student grouping. (Some people call this 'changing the texture' of the lesson.) By following this principle, a one-hour class may begin with listening, move on to speaking, change to reading, and end with writing, or indeed the skills focus may be cyclical. The variety will, at least in theory, prevent the class from stagnating by providing peaks of attention that promote learning and retention.

Another advantage of integrating skills is the way in which different skills work together to reinforce language acquisition, allowing the student to see connections and process language in different channels. For example, a student may write a word (in the writing segment of the lesson), then a few minutes later hear that word (in the listening segment), then say the word (in the speaking segment). Such repeated exposures in different modes are an essential feature of language acquisition.

What are the advantages of a 'listening only' approach?

The main advantage of a 'listening-only' course is its intensity. The students come to class prepared to tackle just the skill of listening and they learn, through sustained focus and repetition, to employ a number of listening strategies. The continuity of a listening-only course also allows students to make progress – and monitor their progress – in this skill more easily.

Whichever approach is employed, we believe that active listening can benefit from the use of all skills. The majority of the activities in Part II include either writing, reading or speaking, as well as listening. As we see it, the use of other skills brings a sense of completeness to the activities, enabling beneficial aspects such as **schema activation**, memory enhancement, task completion and personalised follow-up.

How can we combine active listening with the other skills?

Listening and speaking

Speaking can come before or after listening, and in communicative classrooms it often happens at both times. We speak before listening in order to activate the schemata: we might discuss the subject of the listening, describe a picture related to it, or brainstorm ideas on the topic. We speak after listening in order to compare and confirm our ideas, question the content, identify 'trouble spots', or extend the task in some way, e.g. complete a story, exchange information, solve a problem. One of the central arguments for integrating listening and speaking is, of course, that they co-occur so frequently 'in the real world'. A good conversation is a dialogue involving listening and speaking on the part of both participants.

Listening and reading

The students may read a short passage in order to understand background information and preview new vocabulary before they listen. Other activities that involve reading before listening include: read a summary that contains four content errors (then listen to find the errors); read a gapped summary (and listen in order to complete it); read statements and say if you think they are true or false (listen to find out). Students may read after listening in order to confirm their ideas (e.g. they read the transcript) or to extend their knowledge of the content of the passage. As with listening and speaking, outside the classroom listening and reading are often undertaken together such as when we read subtitles while we watch a film or take in information on slides while we watch a presentation.

Listening and writing

Before listening, the students can be given the topic of the passage and asked to 'free write' – a technique designed to generate ideas and text on any given subject – or write 10 words they think they will hear. While listening and after listening, students are often asked to take notes, write answers to questions, or write summaries. Other post listening activities involving writing may be more creative: write an alternative ending, re-imagine the content in a different genre (e.g. write it up as a scene from a film), or write a personal reflection on the issue. Real-world applications of these two skills combined include note-making from lectures, taking dictation, writing the minutes of a meeting, or scribbling down directions or instructions.

Overall, practising two or more skills together in the classroom makes perfect sense when we use them together so frequently in the real world. Here are some activities from Part II that integrate the skills:

Skills	Activities	When is other skill used?	How is other skill used?
Listening and speaking	Top 10 List	Before listening	Anticipate tips
	My Turn/Your Turn	After listening	Take turns to say what they understood about the passage
Listening and reading	Split Notes	Before listening	Read key words from the passage
	Transcripts	While/after listening	Read to confirm what they heard
Listening and writing	KWL	Before listening	Write what they know and want to know about the topic
	Finish the Story	After listening	Write the ending of the story

Adapting the listening curriculum: institutional constraints and opportunities

How can a listening curriculum be adapted for schools with little equipment?

Many schools do not have advanced equipment, and some have no equipment at all. In these cases, the teacher's voice becomes the main source of listening for the student. And not only the voice: the teacher needs to be able to adopt various personae as he or she reads scripts, embarks on mini lectures, tells anecdotes. Even with this limited exposure to the spoken language, students can still develop their listening with the help of a flexible, creative teacher.

The other main source of listening input in schools with no equipment is the students themselves. The following activities can be done easily using just the teacher's or students' voices (and in some cases a few simple objects or bits of paper): *Guided Journey* (see Affective Frame); *Memories* (see Top Down Frame); *Interactive Quiz* (see Interactive Frame); *Listening Circles* (see Affective Frame); *False Anecdote* (poster paper – see Top Down Frame); *Photo Album* (photos – see Affective Frame); *Guiding Objects* (personal objects – see Top Down Frame); *Race to the Wall* (labels for the walls – see Bottom Up Frame); *Milestones* (Post-It® Notes – see Interactive Frame).

How can an active listening curriculum be adapted for schools with tightly controlled curricula?

Some schools impose a tightly controlled curriculum that asks teachers to teach specific lessons at specific times using set materials. When the curriculum is tightly controlled, teachers might wish to use shortened versions of the activities in Part II as lead-in activities. Some of these activities lend themselves well to this because they do not require a student-generated speaking outcome. We have used *Interactive Quiz* (see Interactive Frame), *Guided Journey* (see Affective Frame), and *Race to the Wall* (see Bottom Up

Frame) as warmers for the beginning of classes, each taking no more than a few minutes and each serving as a way in to the topic of the day.

How can an active listening curriculum be adapted for different institutional philosophies?

What if the institutional philosophy is based on a 'heads-down' approach that is at odds with the very notion of active listening? Teachers can still promote active listening, but the *action* will be mental rather than physical. While the students may appear inert, by using challenging and motivating activities the teacher can ensure they are deeply engaged in listening. These activities might include an element of problem solving (*The Right Thing*, see Top Down Frame), creativity (*Finish the Story*, see Affective Frame), a memory challenge (*Total Recall*, see Bottom Up Frame), or a cognitive challenge (*Map Readers*, see Bottom Up Frame).

How can an active listening curriculum be adapted to make the most of opportunities provided by the institutional context?

Besides institutional constraints, there are institutional and situational opportunities: in ESL contexts, students may have access to large numbers of native speakers as well as the media of the host culture – radio and TV – that can provide ample opportunities for student-listeners. We encourage students to take advantage of such opportunities. Any interaction with a host culture can be a learning opportunity, and this is particularly true when it comes to listening. We have taken students to numerous places in the host community to practise the skill: (1) to art galleries and got them to use the audio guide as listening material; (2) to Speakers' Corner in London's Hyde Park (a place where members of the public are invited to stand on a pedestal and make speeches, often concerning issues of social justice) and had the students listen for gist and detail; (3) to comedy clubs, and judged the success of the students' listening by their laughter; (4) to the cinema; (5) to readings by authors; (6) to sports clubs; (7) to cafes, bars and restaurants.

In EFL contexts, there are often international clubs and groups – in sports, recreation, special interests – that students can access to learn English as an international language. Most cities in the world today have English-speaking groups of various kinds that students can visit or join for listening and interaction opportunities. Furthermore, in technologically well-equipped institutions (e.g. with suites of laptops or tablets and broadband), the students can also benefit from online chat areas and games.

All of these provide real-life listening experiences, some interactive (cafés and restaurants), some extended (cinema, Speakers' Corner), and all motivating for inquisitive students. Of the activities from Part II, *Vox Pops* (see Autonomous Frame), *My Listening Library* (see Autonomous Frame), and *Guest Speaker* (see Interactive Frame) are particularly relevant in EFL contexts.

How can a listening curriculum be adapted for very large classes?

The first issue in very large classes concerns acoustics. Microphones and amplifiers may be necessary and, if so, they need to be well-positioned. It may be beneficial to arrange seats around available speakers rather than in a traditional seating set up. If using video, teachers should check that the size and position of the video screen allows access by all students.

The second issue concerns space. Can the students see the teacher clearly? Listening involves understanding body language and lip reading (that's one reason why phone conversations in the target language are so difficult for students) and if they cannot see the teacher clearly this will be a hindrance to their performance as listeners. Are the students in groupings that are conducive to listening to one another? Chairs arranged in rows impose limitations on group work because the students need to turn their head in order to see only their immediate neighbour. Can the chairs be arranged in semi-circles in order to promote student-to-student communication and eye contact?

The third issue concerns pedagogy. In very large classes, the temptation is to teach the students in lockstep (everyone doing the same thing at the same time). Is it possible and is it preferable to have students work in pairs and groups when there are 100 people in the room? Our experience suggests it is. Lecturing is effective for giving general messages and effecting attitude shifts, but it is a far less cogent form of instruction for working with specific skills, so we put students into small groups to do hands-on, collaborative work. This small group format is effective even in very large classes. A number of the activities in Part II can be adapted for very large classes, for example: *Listening Circles*, *Finish the Story* and *Guided Journey* (all in the Affective Frame); *Keep Doodling* (see Top Down Frame); *Guest*

Speaker, *Milestones* and *My Turn/Your Turn* (see Interactive Frame); *KWL Chart* and *Split Notes* (see Top Down Frame).

How can a listening curriculum be adapted for very small classes?

In classes of just three or four students, it is often preferable to have everyone work together. Sometimes, if the activity requires pairwork but there is an odd number of students, the teacher will partner up with a student and step into a role. When the teacher is taking a role in pair work, it is difficult for him or her to monitor the other students. When it is necessary for the teacher to act as a participant, it is best to ensure that you do not always work with the same student. An alternative is to use the non-participating student as an observer who later gives feedback, or that student can simply wait their turn and then replace a participant at an appropriate stage of the activity.

With one-to-one lessons, the teacher takes an active role as a conversation partner. In this scenario, many of the activities in the Interactive Frame come into play, particularly those that require an element of reciprocity and turn-taking. We have found that the following can be adapted easily to one-to-one lessons: *Blind Forgery*, *Milestones*, *My Turn/Your Turn* and *Interrupted Story* (see Interactive Frame); *False Anecdote* and *Memories* (see Top Down Frame).

Adapting the listening curriculum: different student populations

How can a listening curriculum be adapted for students of different ages?

The age of the students may have a big impact on their listening ability. Young children, and some older children, are famously incapable of listening to anything for more than a minute or two unless it is a vividly told story (in which case they usually pay rapt attention). The key is either to get young children hooked on a story or to get them moving. Activities such as *Fly Swatter* (see Affective Frame), *Word Grab* and *Race to the Wall* (see Bottom Up Frame), which use action, tend to work well. And any activity involving TPR (total physical response), such as *Action Skits* (see Bottom Up Frame), is also likely to be popular. While there is always the danger of complete chaos with such activities, more experienced teachers will use their classroom management techniques to keep the students on task.

Older learners – those in their 70s or above – may find some of the more physical activities challenging, but, unless there is a limiting disability, we encourage the full range of listening activities, including the kinaesthetic ones, for learners of all ages.

How can a listening curriculum be adapted for an EAP class?

EAP (English for Academic Purposes) classes are appropriate venues for lecture input. Lecturing is still the main form of teaching in global higher education, and the students need practice in understanding complex ideas delivered in condensed lecture formats. These formats have, in recent years, become increasingly multimodal, with PowerPoint presentations and other visual and textual information accompanying auditory content. For the students to gain practice in dealing with this type of input, long lectures divided into shorter chunks, with accompanying tasks, are appropriate. Using material with an academic slant, the following activities can easily be adapted for EAP students: *My Turn/Your Turn* and *Pecha-Kucha* (see Interactive Frame); *Split Notes* (see Top Down Frame); *Details, Details* (see Bottom Up Frame); *Good Question* (see Top Down Frame); and *Guest Speaker* (see Interactive Frame).

How can a listening curriculum be adapted for a Business English class?

With Business English classes, the focus tends to be on interactive listening. Many of the pursuits of businesspeople – negotiating, going to presentations, attending meetings – are based on highly attentive listening followed by interaction. The use of strategies is essential, particularly conversational repair and asking for clarification. Business English teachers know that any business listening syllabus involves the teaching of dozens of strategic phrases through which listeners may regain control of the conversation. In short, the business *listening* syllabus is, in part, a business *speaking* syllabus.

The type of activities that lend themselves well to adaptation for Business English students include: *Pecha-Kucha, Paraphrase* and *Guest Speaker* (see Interactive Frame); *The Right Thing* (see Top Down Frame); and *Shadowing* (see Bottom Up Frame). For these, the teacher needs to find material with a business slant. For example, *The Right Thing* requires a business ethics dilemma rather than a general dilemma.

How can a listening curriculum be adapted for an exam class?

Some of the main features of listening exams are the abilities to infer, to note down answers rapidly, to hold detailed information in memory while dealing with other incoming information, and to cope with 'divided attention', e.g. reading and listening simultaneously. Teachers of exam classes need to be aware of these features of exams and to practise them accordingly with their students. For exam-focused classes, explicit strategy training is invaluable. The strategies will include reading questions before listening, listening for key words and synonyms, anticipating information, and making informed guesses. Here are some activities that work well with exam classes: *Total Recall, Details, Details* and *Pause and Predict* (see Bottom Up Frame); and *Split Notes* (see Top Down Frame).

How can listening activities be adapted for testing purposes?

All of the listening activities in Part II are designed for teaching purposes, and the main criterion for success is active engagement. Most of these activities can be used for additional assessment if you define a clear, valid criterion for success in the activity and evaluate accordingly.

If we want a listening activity to be used as a tool for assessment there are two principles we need to follow: **validity** and **reliability**. A valid listening test should ask the student to demonstrate comprehension, usually through answering questions, *and do nothing more*. A valid test measures only what it is intended to measure, and therefore it needs to focus on the part of the task that is most concerned with listening for understanding. This means that for a task to be valid as an assessment of listening, we need to minimise the use of productive skills (writing and speaking) during the task and avoid focusing on them as part of the assessment. Otherwise, students may be penalised for weaknesses in these, even if they understood the passage well.

Test reliability means that a test produces consistent results over a period of time, with different participants, and in different conditions. It should make no difference to the results, for example, who grades the test, what mood they are in, or whether the test is taken in the morning or afternoon.

Below we examine the efficacy of some listening activities for testing purposes.

Comprehension questions

Advantages: you can focus on the most important aspects of the passage.

Disadvantages: if the questions are written, you are also testing the student's reading skill and, if responses are written, you are also testing writing skill. With comprehension questions, there is also the difficulty of finding a balance between global questions (that ask students to infer cause-and-effect relationships and synthesise information) and local questions (that ask for specific details).

Multiple choice questions

Advantages: you can achieve good reliability (because there is little variation in the type of input and output). There is no use of productive skills, meaning that speaking and writing are not tested. Multiple choice questions are easy to grade.

Disadvantages: students can make guesses. It can be difficult to write distractors that don't 'derail' listeners.

Dictation

Advantages: dictation is reliable inasmuch as it is clear what is right and wrong.

Disadvantages: there is not much real world applicability. Dictation tests writing skill, so it is not a pure listening test. It also tends to be tedious to administer, take and score.

Summary writing

Advantages: it is clear whether the listener has understood the passage or not.

Disadvantages: it may be necessary to teach students the skill of summary writing beforehand. Summary writing tests memory; students may under-stand everything but be unable to remember it all when they need to write.

Here are some activities from Part II that lend themselves directly to adapta-tion for tests: *Total Recall* (see Bottom Up Frame); *Split Notes* and *Good Question* (see Top Down Frame); *Wrong Words* (see Affective Frame).

How can listening activities be adapted for different levels?

The students' level is clearly a major factor in classroom tasks. As their level of English rises, students are able to cope with more difficult passages or dialogues. The concept of 'difficulty' may be measured in numerous ways:

Speech rates

The faster a speaker talks, the more difficult it is for a listener. Real-time processing of auditory input is harder when the listener has less time. Also, many of the features of fast speech, such as elision, assimilation and vowel reduction, make word recognition more difficult for listeners.

Accent familiarity

The less familiar the accents of the speakers, the more difficult it is for the listener to understand. Typically, the problem for listeners is that speakers with regional accents pronounce vowel sounds differently from what the listener is used to hearing. This may render words unrecognisable.

Length

Lower levels struggle to follow extended passages. While listening in real time, students sometimes miss incoming auditory information because they are still processing the input they just heard. This problem is exacerbated with longer passages.

Number of speakers

The more speakers, the more difficult the conversation is to follow. Particularly with audio-only recorded material, it can be tricky for listeners to know who is speaking, and thus they lose track of the thread of the conversation.

Linearity

Lengthy digressions and flashbacks are harder to grasp than straight narrative. It is easier to understand a narrative in which the order of telling reflects the order of events as they happened.

Discourse structure

Some discourse structures (sometimes known as **formal schema**) require deep cognitive processing, while others are more straightforward. For example, a narrative structure is easier to follow than a more abstract *cause-effect* structure.

Language level

The use of low-frequency words and phrases, subject-specific jargon, or colloquial idioms makes listening more difficult. Every time students hear unknown words, they need to guess the meaning in real time, leaving them with less 'cognitive space' to continue processing incoming information.

Density

Many ideas arriving in close proximity makes it harder for listeners to follow, especially if there is little redundancy and repetition. Also, passages with numerous complex and embedded sentences will be more difficult for listeners.

The following table shows what activity types beginning, intermediate and advanced students may find most effective, along with targeted skills and strategies. Naturally, this table is not exhaustive; it is provided as a simplified guide to the type of activity and teaching focus that may be appropriate for the different levels.

Level	Skills they most need to improve as listeners	Example activities
Beginner	1. Discriminate between phonemes	• *Word Grab (variation)*
	2. Recognise words in connected speech	• *Race to the Wall*
	3. Understand gist	• *KWL*
	4. Match words and sentences to pictures	• *Photo Album*
	5. Follow instructions	• *Action Skits*
	6. Follow a basic storyline	• *Memories*
	7. Recall and repeat key words and phrases	• *Whisper Dictation*
	8. Show that you are following a conversation	• *Cloud Discussions*
Intermediate	1. Improve capacity to recognise words in fast connected speech	• *What's the Line?*
	2. Understand main ideas and supporting details	• *2-20-2 Pictures*
	3. Make inferences concerning topic and speaker	• *False Anecdote*
	4. Evaluate a speaker's content	• *Top 10 List*
	5. Follow a story that has some digressions	• *Finish the Story*
	6. Provide backchannelling to the speaker	• *Blind Forgery*
	7. Show basic recall and ability to summarise short passages and conversations	• *My Turn/Your Turn*
	8. Listen to authentic extracts outside of class and provide short summaries	• *Learn Something New*
Advanced	1. Use sentence stress and intonation to detect nuances of meaning	• *Emotional Scenes*
	2. Recognise discourse features of long turns/ lectures	• *Keep Doodling*
	3. Listen critically to evaluate complex or extended content	• *Film Review*
	4. Use background knowledge to infer meaning from incomplete aural signals	• *Pause and Predict*
	5. Evaluate different perspectives on a topic	• *Webquest*
	6. Analyse the thinking processes of the speaker	• *Listening Circles*
	7. Respond to the speaker in ways that deepen the conversation	• *Guest Speaker*
	8. Listen to authentic extracts on your own and provide meaningful summaries and critiques	• *My Listening Library*

There are three main approaches to adapting a listening curriculum for lower levels: (1) change the input to make it easier (use different recordings);

(2) simplify the task; or (3) change the task procedure. Examples of the latter might include doing extensive pre-listening work so that the students are primed to succeed, providing scaffolds such as peer support and transcripts, and playing recordings several times in small chunks.

To adapt the programme for higher levels we do the opposite: use harder recordings, make the tasks more challenging, and remove some of the support structures that scaffold the listening experience. One common support structure is the pre-teaching of difficult vocabulary. Teachers may choose, for higher-level students, not to pre-teach vocabulary, instead asking the students to guess the meanings of unknown vocabulary from the context. Another form of scaffolding is previewing the content of the listening via discussions, images, texts, etc. This process, too, can be abandoned if the teacher wants to increase the challenge for the students.

How can a listening activity and a listening curriculum be adapted to cater for students with low motivation?

Teachers in this situation need to get to the root of the issue: why are the students unmotivated? It may be for any number of reasons: the class is at the wrong time of day for their body clocks, they lack aptitude for language learning, they don't understand the wider importance of the class, the configuration of the room induces disengagement, there is a mismatch between teaching and learning styles, the students' expectations are not be being met, or the class doesn't speak to the students' sense of self.

Then what can teachers do? They personalise the topics to make them relevant to the students. They try to provide an element of student choice, because when people make a personal investment, their motivation increases. Teachers experiment with a number of features of the class: the physical set-up of the classroom, the groupings and pairs that work together, the order in which activities take place, and the type of content. They also try to use engaging activities regularly. Here are some activities, mainly from the Affective Frame in Part II, that have helped motivate our students: *Punchline, Wrong Words, Pinch and Ouch, Fly Swatter, A New Skill, Pecha-Kucha, Whisper Dictation, Guest Speaker, Milestones.*

Trying to elicit choices for an activity and 'debriefing' students after an activity can also help to 'troubleshoot' motivation issues. For example, posing easy-to-answer questions like, 'Which of these two songs would you like to use for the next class?' or 'Which part of the activity was most interesting?' can help engage students in the learning process and begin to awaken motivation.

How can a listening activity be adapted to cater for students who do not have an auditory learning style?

Learning styles refers to the way in which people prefer to take in new information. Researchers have identified many learning styles, some of which are included in the following table, along with brief descriptions and a list of the type of activities that typically appeal to these students. The learning styles in this table are particularly associated with Howard Gardner's *Theory of Multiple Intelligences* (Gardner, 1983; 2011).

Learning style	Description	Appropriate listening activities (from Part II)
Visual (Spatial)	• Learns best by *looking* • Likes to use pictures and images • Has good spatial awareness	• *2-20-2 Pictures* • *Blind Forgery* • *Photoshop* • *Keep Doodling*
Auditory (Musical)	• Learns best by *listening* • Enjoys songs and other aural input	• *Guided Journey* • *Wrong Words* • *Interrupted Story* • *Guest Speaker*
Kinaesthetic	• Learns best by *doing* • Likes physical activity and hands-on learning	• *Word Grab* • *A New Skill* • *Fly Swatter* • *Race to the Wall*
Verbal (Linguistic)	• Learns best by *reading and following information* • Makes sense of the world through language	• *Webquest* • *What's the Line?* • *Shadowing* • *Pecha-Kucha*
Logical/mathematical	• Learns best by *using logic and reasoning* to work out systems	• *Learn Something New* • *KWL* • *Pause and Predict* • *False Anecdote*
Interpersonal	• Learns best by *working with others* • Enjoys group and pair work	• *Photo Album* • *Milestones* • *Paraphrase* • *Conversation Corner*
Intrapersonal	• Learns best by *working alone* • Enjoys self-study and understands own motivation	• *Listening Games* • *News Hound* • *Transcripts* • *My Listening Library*

In reality, people do not always fit neatly into the categories in this table. Many of us fit several descriptions, and our learning styles may vary from day to day or topic to topic. While some teachers go to great lengths to categorise their students' learning styles, we feel it is sufficient to be aware of the differences in the classroom and provide plenty of variety in order to cater to those differences.

Students who may be very weak in auditory learning, as a preferred learning style, *can still become proficient listeners* if they can engage in other types of input processing in conjunction with listening. For example, they may benefit from visual information such as illustrations, diagrams or charts while they listen.

Adapting and creating the listening curriculum: different teachers

How can a listening curriculum be modified by experienced teachers to better fit their class's needs?

Many experienced teachers have an instinct for adapting material to their students' needs. They understand the principles behind the activity and realise that it is only a change of emphasis or a recording with a slightly different slant that is required to make an activity relevant and appropriately challenging to their students. They mix and match, incorporating elements of other activities, leave out certain stages of an activity, extend or reduce other stages, change the groupings, or abandon the activity once it has served its purpose.

Just as they do this with an individual activity, they also do it with a whole curriculum. Following extensive diagnostic work early in the course, the teacher then focuses on improving their students' weak points. For example, if the students struggle with fast, connected speech, the teacher tailors the curriculum so that it contains more news broadcasts or authentic film clips. If the students struggle with listening in real-time face-to-face conversation, the teacher increases the number of interactive tasks. As such, the listening curriculum, in the hands of an experienced teacher, becomes a menu or a set of options to be selected from rather than a set of laws written in stone.

How can listening activities and a listening curriculum be adapted by and for novice teachers?

Novice teachers need to learn how to recognise the principles behind activities. This enables the adaptation of the activity in a principled manner for the teacher's particular context. If you are a novice teacher there are several techniques that will help you: review the aim of the activity, focus on students becoming engaged with this aim, visualise how the entire activity will work before you begin, and then try to avoid dwelling too long on any one step of the activity. Later on, after a period of reflection, try the activity again, making modifications where necessary (e.g. to the timing or the way you group students), and compare the results.

In terms of the listening curriculum, you should try to see this type of 'activity recycling' as a progression. Experienced teachers understand how the listening task on Monday relates to the listening task on Tuesday. In other words, focus on how the curriculum gradually builds on what has come before, and how listening strategies need to be used again and again before they become part of the student's identity as a listener.

How can a listening curriculum be adapted to take into account the different beliefs of teachers?

We all tend to choose or adapt our teaching approaches to fit with our cultural beliefs: our views of individualism versus collectivism, our sense of intimacy versus social distance, our views on the role of the teacher and student in society. Many of these beliefs we acquired through life experience; others we inherited by the ways we were taught and trained.

Some teachers will balk at sharing their personal information (as in the activity *Photo Album*, Affective Frame) because this closes the distance between themselves (the authority figures) and the students (the novices). Others will object to 'telling lies' (as in *False Anecdote*, Top Down Frame), even for a pedagogical game-like activity. Some teachers may view activities that involve physical action (*Race to the Wall* and *Word Grab*, Bottom Up Frame) as childish and an affront to the dignity of adults. Yet others will decry the deliberate distortion or disruption of the input (*Whisper Dictation* and *Interrupted Story*, Interactive Frame) as unnatural and inauthentic. Our advice concerning activities is for teachers to consider the aims of the activity and find alternate ways to achieve these aims or, if the aims are not appropriate for your students, omit the activity.

On a more general note, we believe that teachers should recognise that there are many ways to learn and to teach. By observing other teachers, watching lessons online, reading widely, and seeking opportunities for professional development, teachers gradually come to realise that other classrooms differ from their own. Sometimes it is not the curriculum that needs adapting, but the teacher who needs to adapt. Part of our professional development is testing our comfort zones, and experimenting with new styles of teaching, as well as encouraging our students to try new styles of learning.

How can teachers adapt activities to encourage active listening?

In order to adapt activities, teachers need to have a deep and conscious understanding of what goes on in their classrooms and why. They need to become adept questioners, note-takers and 'noticers'. Which students work well together? What is the optimum group size in the class? How long can an activity go on for? Is the classroom set-up conducive to physical movement? Can all of the students hear the recordings? What outside resources do the students have access to?

Experienced teachers may be able to answer these questions intuitively. Others will need to develop conscious reflection practices, such as making notes in a teaching journal, that help them to answer these questions. Good teachers gradually become experts at recognising the unique characteristics of any one class, and this knowledge is what allows them to adapt activities successfully.

Below are some practical ideas for adapting activities to encourage active listening. Tell the students they will need to:

- collaborate with other students to write a version of the passage afterwards;
- write about the passage afterwards (a summary, an opinion piece, a news report);
- describe five things they learned from the passage;
- raise their hand every time they hear a certain word or phrase;
- mime the actions and events contained in the passage as they listen;
- correct five factual mistakes in a written summary of the passage;
- research the topic for homework;
- tick whichever of their predictions about the passage are correct;
- make a visual representation of the passage;
- devise and act out a skit of the passage;
- give advice or solve a problem contained in the passage;
- give a talk based on the model passage.

How can teachers create their own activities to encourage active listening, i.e. what are the principles of these activities?

As you reflect on the value of each activity you use in class, and add your own imagination, you will begin to want to create your own activities. This is something we genuinely encourage! Creating our own activities is one way in which we contribute to the wider profession; the activities may appear in journals or they may form the bases of conference presentations. Above all, creating activities gives you a sense that you are developing professionally.

In order to create effective activities, we need to understand the principles behind what we're doing. Once we are sure of the principles, all new activities are honed through a process of trial and error. We have the spark of an idea that we mentally evaluate. If it seems worth pursuing, we bring it to class. We tentatively try it out with our students. We then re-evaluate the activity. What worked? What didn't work and why? How can it be improved? We then retry, perhaps with another group of students. Again, we use our critical faculties to evaluate the activity. And the process continues until we feel happy with it. It then becomes a part of our identity as teachers.

What about activities specifically pertaining to active listening? Here we need to visit the research frames from Part I and extrapolate the principles of the activities presented in Part II. These frames and principles will guide teachers in the creation of activities. See the Frame–Principles table below for a summary.

Though the frames and principles can provide guidelines for developing new listening activities, the real work requires creativity on your part. By this we do not mean creativity in the sense of making masterpieces; instead, we see it as linked to problem-solving. There is a problem or a puzzle (e.g. the students are poor listeners) which needs a solution. So the teacher tries out something new. That 'something new' constitutes creativity, using your imagination and your own ideas.

Often, the creative aspect consists of taking an activity from another discipline and adapting it to your classroom. This is how *Pecha-Kucha* (see Interactive Frame) came to be used in language classrooms, having begun in an architectural firm. *Bucket List Bingo* (see Bottom Up Frame) is based on a popular game played for money in many countries, here used in the language classroom. Visualisation (such as *Guided Journey* in the Affective Frame) has long been used by therapists and sports coaches; here it gets a twist and becomes part of the language teacher's repertoire.

Frame	Principles
Affective Frame	Use a playful approach: • Find a built-in stimulus that motivates the students • Use personalisation as a way to involve students in the content • Give students tangible actions to perform • Relieve the stress inherent in real-time listening through a sense of *play*
Top Down Frame	Use an ideas-building approach: • Provide a pre-listening stimulus • Use the students' prior knowledge • Focus on student questioning and ways to generate questions • Promote tolerance of delayed confirmation of answers • Get students to provide their own reason to listen
Bottom Up Frame	Use a language-noticing approach: • Focus the students on limited listening goals involving small details • Direct attention to the building blocks of language, not just the building • Include a detailed post-listening session involving small units of language (e.g. phonemes) • Diagnose what was hard to understand and why
Interaction Frame	Use a collaborative approach: • Set up a two-way conversation, even if one party is dominant • Create a gap in understanding between the two (or more) parties in the conversation • Teach discourse phrases for regaining control of a conversation
Autonomous Frame	Use an independence-building approach: • Find and use authentic sources • Encourage student independence: students making choices out of personal interest • Help students develop strategies for coping with above-level material without teacher support

What ideas and creative mindset can help us to devise new activities?

As teachers, we are continuously faced with the challenge of needing something new and fresh to inject into the classroom. Our experience has taught us that successful teaching often comes from a creative mindset rather than an abundance of resources. Here are some of the ideas that have been fruitful for us:

• Cross-fertilise ideas by borrowing from other fields: sports, drama, children's games, advertising, marketing, art, science, politics.

- Borrow from popular culture: talent shows (*Pop Idol*), reality shows (*Big Brother*), or quiz shows (*Who Wants to be a Millionaire?*).

- Visualise the classroom space in ways that you do not normally: see every physical part of the room as an opportunity for movement, display and student ownership.

- Think in multi-dimensions: activities can involve a variety of objects, voices, media, images, movement and competition.

- Consider what would surprise your students the moment they walked into the classroom.

- Think of the different senses: sight, sound, smell, touch, taste. Which do we *not* normally use in English classes, and how might these be used?

- Consider unpredictable outcomes. Normally, classroom procedures follow predictable patterns and end in predictable ways. What if no one knew how an activity would end?

- Think about how the students can use their imagination in class to adapt or bend the rules of normal classroom behaviour.

- Brainstorm ideas based on a set of verbs: create, design, find, make, imagine, write, extend, use, bring, tell.

The key, when devising activities, is to link creativity to context. The most brilliant boat design will be no use if you live 5,000 kilometres from water. True innovation is always linked to concrete purposes.

Part IV
From Implementation to Research

Part IV presents the final stage of the research–practice–research cycle and is divided into two sections. The first section is a guide for conducting action research in the area of listening, based on the active listening frameworks used in the previous parts. The second section presents research links organised by topic that offer examples of useful queries to guide action research.

Conducting action research

For most teachers, deciding to launch into explorations of issues involving our own students is a pivotal stage in professional development. By entering into such exploration, you can: (1) improve the quality of your teaching; (2) address problems in the classroom in a principled way; and (3) contribute to research into classroom practice in the field of active listening and language acquisition.

This kind of practice-focused research is usually referred to as **action research**. In a sense, it is the 'active teaching' parallel to the 'active listening' framework that we have developed in this text. Just as active listening is based on engagement, action research is based on an intention by the teacher to develop an idea, explore an issue, or address a problem.

What is action research?

Action research is a form of **applied research** which aims to develop knowledge that can be *used directly by practitioners*. It is essentially reflecting on new, interesting or problematic areas in your teaching. Action research differs from basic research in that the purpose of basic research is to understand a process or phenomenon without the necessity to apply the results. Action research also differs from simple reflection in that it is undertaken in a planned and structured way and produces evidence in the form of observable data. This data can then be used to influence or direct changes in, for example, classroom procedures or modes and types of learning.

Many teachers already do an implicit form of action research, continually gauging a sense of their own 'flow' in teaching, and making adjustments accordingly (Tardy and Snyder, 2004; Burns and Richards, 2009). Although such implicit research can be satisfying, taking the step of documenting procedures and articulating the results can have additional personal and professional benefits (Wright and Bolitho, 2009; Yang, 2009).

Many teachers initially balk at the idea of doing research because of time constraints, but action research projects need not be very time-consuming. Some projects can be as short as a day or two, involving simple observations, interviews and reflections.

Why do action research? Where to start?

The main reason to do action research is to address a problem or issue in your practice, perhaps one that has been lingering for some time. The most common way to start an action research project is simply to *formulate a question* that you would like to answer. Some educators refer to this as 'problematising' a situation, though it is important to note that there is no connotation of teacher or learner deficiency in problematising a learning situation (Loughran, 2010). Problematising in this sense simply means *working out and articulating a puzzle* that has presented itself, and then working towards an understanding of the puzzle and possible ways to solve it.

You probably have your own teaching-learning questions in the field of active listening waiting to be formulated, but here are some that we have heard.

Affective issues

- My students don't seem motivated during listening activities. Why not?

- One student in particular does not engage in listening activities in my classroom. I wonder what's going on?

- Some students in my class seem to be trying hard, but are making no discernable progress in listening. What can I do?

- I have a few students who seem to give up easily during lectures. They take minimal notes and just give perfunctory answers on quizzes and tests. I'd like to give some advice but, beyond simple encouragement, I don't know what to do.

- I have a few students who just go through the motions during tasks. I think they believe they're going to pass the class and graduate, but I know they're not ready. What should I do?

Top down issues

- How often should I play a recording?

- How do I know when a listening extract is just too difficult for my students? I know how to make easy tasks, but it doesn't seem right to make a simple task when the listening itself is so complex. I want to use authentic materials, but I think most are just too difficult.

- When should I stop the recording? I can tell some students get confused after a little while, and I'm tempted to stop and replay, or stop and ask questions.

- I tell my students to use their background knowledge to figure out things when they listen, but I don't really know how to do this. Just telling the students to 'think about it' doesn't help. What can I do?

- I know that a lot of what speakers mean has to be inferred – the students have to learn to listen between the lines. Can this be taught?

- I'm not sure if my students are recognising discourse markers (*First of all, Another thing is, Finally,*) or cohesion markers (*such as,* relative pronouns, pronouns) when they listen. I think these bits happen so fast that they just don't notice them. What can I do?

- I have some students who refuse to take notes. They don't think it helps them listen. I hate to give them poor marks for the class, but it is a 'listening and note-taking' class. What's a fair thing to do?

- Should I allow students to take notes during a listening activity?

- I have one student who tries to write down every word during a lecture. I tell her just to focus on main ideas, but she's not sure what I mean.

Bottom up issues

- My students seem to panic when they hear a word or phrase they don't know. I tell them not to worry, but they do worry. They want me to stop and explain everything they don't know.

- Is listen-and-repeat a useful listening exercise? My students love it, but I'm not sure it's considered useful anymore.

- I wonder if I need to teach more about intonation. I really don't know where to begin.

- Does reading aloud help students with listening?

- Do students need to be able to pronounce all of the reductions and assimilations I'm teaching, or do they just need to be able to recognise them?

- A few students are preoccupied with getting every word when they listen in class. They seem to be missing the main point all the time, and just ask questions about grammar and vocabulary. Should I modify my lessons to give them what they want?

- Many of my students say they have trouble listening because English speakers simply speak too fast. They can't control the pace of the speaker. What can I do?

- My students are baffled by English phonology, especially reductions and assimilations. Is there any way to teach phonology systematically?

- A lot of my students don't know common idioms, so they get confused when they hear colloquialisms.

Interactive issues

- When I use interactive listening and speaking tasks, a few students just give the easiest responses, usually 'I don't know'. How can I get them to take a more proactive role?

- Some of my students have real difficulties when talking with native speakers. They do fine in class and with me, their teacher (a native speaker), but not with other native speakers. How can I help them?

- Are presentations useful for classroom listening practice? My students seem to enjoy preparing and giving presentations, but the audience tends to get bored as the presentations go on too long.

- Some students in my class simply won't do any autonomous listening outside of class. They don't respond to my exhortations. I don't know what to do.

- Most of my students speak imperfect English, so I'm worried about having them do pair work together. What can I do?

- My students seem animated during pair activities, but I notice that they often speak in their L1 when I'm not there. Is this something I should be concerned about?

Autonomous issues

- I have too much to do already with classroom teaching, so I'm not inclined to spend even more time setting up a self-access listening programme. Is there any way to do this without adding more time to my schedule?

- Is it better to assign out-of-class listening to things I've already listened to myself, or is it OK to assign things that I haven't heard?

- I know my students love music, but I think I'm out of touch with what they like to listen to. How can I recommend popular music that is useful for their learning?

- What are listening strategies actually? I hear talk of meta-cognitive, cognitive and social listening strategies. How are they different? Should I be teaching these to my students?

- Is reading and listening (like using graded readers with audio recordings) together a good idea?

- Should I allow my students to use subtitles, either in English or in their L1? It seems like it's not a real listening activity if they're depending on the script.

- I know there are a lot of great resources on the internet, but I'm worried about students accessing all kinds of nasty stuff if I give them internet-based assignments. What can I do?

For all of these questions, there probably are prescribed answers, formulas that have been given by 'experts'. Many of these issues have been addressed directly or indirectly in this book. While it is important to stay abreast of what our most informed colleagues are trying, it is also important to look to ourselves to find the solutions that fit best in our situation. As teachers we have come to believe that it is actually an illuminating and rewarding process to find one's *own* answers to questions, even when the available research seems to have provided the best answer already! Being engaged in this process can be one of the most self-renewing activities for a teacher – whether or not the teacher decides eventually to share the findings through a journal article, a conference presentation or a faculty meeting report.

What methods should I use for action research?

Action research generally involves an eclectic combination of methods, utilising only those data-gathering methods that are needed to explore the problem and not likely to disturb the teaching-learning environment. Even though it is impossible to completely avoid what has been called the Observer's Paradox – the phenomenon that any observation will influence the people being observed – there are a couple of unwanted effects that can be avoided: the 'Hawthorne Effect' (where the research focus creates a motivation to perform better during the observation) and the 'John Henry Effect' (where the participants resent the fact that something that has been 'normal' is now being 'problematised'). The most effective action research is done in a way that does not introduce new people or systems (such as experimental settings) into the context and keeps the process transparent to all participants.

Because action research is oriented towards finding out not just what people do, but exploring the relevant thoughts and beliefs of participants in relation to the 'problem' or puzzle, some kind of information gathering is needed. The most commonly used methods are:

Journal

Students keep a scoresheet, a diary or journal (which could include oral journaling with a recording device on a mobile phone or a portable mp3 recorder), making comments in response to a question or directive related to their listening behaviour (e.g. make a note of any time you listen to English outside of class during a 24-hour period).

Questionnaire

Students fill out questionnaires or surveys to investigate areas of interest. This could be a set of situations about learning style preferences, favourite activities for studying listening in class, suggestions for improving the quality of the class, suggestions for out-of-class learning options, or listening strategies that students use. For example, here are some questions from a listening strategy survey:

	I'm not sure	I never do this	I rarely do this	I sometimes do this	I often do this
1. Before starting to listen, I think of what I might know about the story or topic.					
2. I use sound effects and tone of the speaker's voice to help me guess the meaning of words.					
3. As I am listening, I predict what will happen.					

Interview

Students are interviewed by you or by a colleague concerning your focus questions. This might be about their favourite activities for studying listening in class, suggestions for improving the quality of the class, attitudes about correction, etc.

Think aloud

Students produce **think aloud protocols**, in which you guide them through a listening experience, pause at specific points, and ask them what they are thinking or how they are going about trying to understand the listening extract.

Commentary

Students give a snapshot commentary in which you recreate certain moments in a class (if you have video, you can literally freeze frame the moment you want to query) and ask students: 'What is happening here?' or 'What are you thinking about now (in the moment in question)?'

Checklist

Students complete post-activity checklists. At the end of an activity or at the end of a class, you provide a short questionnaire that probes what the students did or learned during the activity or class (e.g. which of these things did you do during the activity? Check all that apply.). You may include questions like: 'I spoke English at least 80 per cent of the time', 'I asked questions when I didn't understand'.

Response

Students complete 'response questionnaires'. After watching a short documentary by a student or teacher describing a learning scenario or strategy or attitude, students indicate the degree to which they 'like' or 'would try' a particular approach.

Try-retry

The teacher does a try-adjust-retry approach with an activity, and then observes the differences. By changing just one variable (for example, type of pre-listening task or grouping of students) you can isolate which decisions contribute to the value of a task. In addition, you may survey the students as to which version of the activity they found more useful.

Test-retest

Test-treatment-retest. This is a simple form of quasi-experimental research – you're using only one group (no comparison group) and no random assignment to groups, so it's not a pure research format. Test your students on some aspect of listening by measuring their performance on a testing task (for example, writing a summary after listening to a short lecture). Then provide your 'treatment', involving teaching of note-taking, reviewing and summarising skills. Test again. If your retest results show some improvement, you have *some* evidence that the 'treatment' may have had a positive effect.

What do you do with the information you collect?

When doing basic research it is important to compare the information you collect and analyse with an external context – what has *already* been studied and published in the same area. When doing applied research, you need to interpret the information you collect in the actual context of the research, including the interpretations you get from the participants themselves.

What the research 'means' to you and others interested or involved depends on a process of interpretation that involves immersion in the data that is produced, as well as obtaining the perspectives of the actual participants. No specific statistical methods are required, though it is useful to compile results if a survey of multiple students or a series of interviews was involved. The outcome of the research is to re-state the problem in terms that allow it to be addressed in the current situation. Essentially, action research allows you to address two key questions: What can we do *now*? What can we plan to do *in the future*? The plan of action then becomes part of a continuing cycle of questioning, experimentation, discovery and change.

How do I publish the findings of my action research?

Action research is designed to address problems arising from practice, so it is not necessary to publish your results or solution to the outside world. Of course, if you feel you have found out something worth sharing, you could consider giving a report at a faculty meeting, writing a short article for a teachers' journal, preparing a presentation for a teachers' conference, or posting your ideas on a blog, either your own or on the blog of someone you follow in the teaching world.

By sharing your findings, or just by posing your questions or posting your ideas so that others may hear them, you are connecting with the teaching community. And in some way you are advancing the professional search for better teaching and learning. As we have said to teachers who attend our teaching seminars, research is a form of contribution, and reading and applying research is a means of developing our profession and ourselves. By using research-based best practices as a foundation for our teaching, we continually advance respect, inclusion, and an enthusiasm for pursuing knowledge.

Research links

As you do action research in your own teaching context, you may wish to consult what other teachers and educators have done. This section includes the Research links provided at the end of the Activities in Part II, grouped according to research topics.

Affective Frame

Kinaesthetic learning

How can I use kinaesthetic (movement-oriented) learning beyond simple commands? Can I incorporate physical actions or more active learning into other kinds of listening activities?

- The kinaesthetic learning style is considered to be one of seven main learning styles, which are 'ways in which we get to know the world'. To explore this concept further, consult the work of Howard Gardner, one of the originators of the learning style concept for education: Gardner, 2011; 1983.

- For an introduction to kinaesthetic learning in a school context, see Shannon Hutton's article on helping kinaesthetic learners succeed: http://www.education.com/magazine/article/kinesthetic_learner/

- For a short, simple introduction to Total Physical Response (TPR), a method that is based on kinaesthetic learning, see Tim Bowen's article on teaching approaches at onestopenglish.com: http://www.onestopenglish.com/support/methodology/teaching-approaches/teaching-approaches-total-physical-response/146503.article

- To explore the foundations of the TPR method, consult the writings of the originator: Asher (2003; 1969; 1966); or visit James Asher's website, which provides an abundance of resources: http://www.tpr-world.com/

- If you would like to look into other comprehension-based approaches that incorporate kinaesthetic learning, see the Natural Approach website: http://naturalway.awardspace.com/index.htm.

Games

Can my students learn listening through games? What kinds of games are most suitable for my students? Can I adapt children's games for older learners?

- Games are widely used as a learning tool in many educational settings, with students of all ages. To explore children's games that are adaptable for language classrooms, see Debra Wise's comprehensive work (Wise, 2003), which contains 450 games that are played indoors and outdoors. Most of these are not directly suitable for language classrooms, but they may serve as an inspiration for adapting and conceiving your own games.

- The notion of 'play' in language learning obviously extends beyond games. 'Play' is now seen to include several categories that apply to children and adults: active play, quiet play, cooperative play, solitary play, manipulative play, creative play, dramatic play. To explore the role of games and play in learning, see an overview article from TEFL games, a learning company based in Thailand: http://www.teflgames.com/why.html

- Some educators employ the term 'serious games' to describe games that include a post-play 'debriefing' phase that is considered valuable for long-term learning. To explore this topic further, see the David Crookall article on simulations (Crookall, 2010).

- Many educators see negotiation as a key ingredient in interactive games for language learning. To explore this topic, see Zheng *et al.* (2009).

- Some language educators view online gaming communities as a vital source of interaction, of a similar value to live interaction in terms of benefit for language acquisition. To explore this concept, see some of Steven Thorne's work in this area, e.g. Thorne *et al.* (2009).

- Many popular games, ones that the students may already know, can be adapted to language learning. For example, to explore the uses of Bingo in the classroom, see Joseph DeVeto's teaching resource site, which gives a quick explanation of the game plus a grid that can be used in class: http://www.teacherjoe.us/TeachersSpeakingBingo.html. If you need a bingo card generator (or other ready-to-to use resources), see the teacher-sharing site http://www.eslactivities.com/bingo.php.

- Games can help in the development of 'emotional intelligence', which has been defined as the ability to monitor one's own and others' feelings and emotions and to use this information to guide one's thinking and actions. To investigate the connections between emotional intelligence and language learning, see Bar-On (1997) and Carroll (1965; 1993).

Drama

How can I use drama or drama techniques to promote listening and speaking skills? Do I have to be a good actor myself to teach with drama?

- Maley and Duff's innovative work on drama techniques outlines numerous exercises that work with scenarios, scripts and short plays – all of these assisting in the development of listening skills. See Maley and Duff (2005).

- Yordana Hristozova has been teaching English through drama techniques for a number of years. She includes 10 principles for effective use of drama, and describes a variety of useful acting exercises for EFL contexts. Her blog on using drama in the classroom is an excellent source of classroom studies on its effectiveness: http://teacherlingo.com/blogs/makeadifference/default.aspx.

Realia

Does realia help students learn languages? If so, how? I'd like to use more realia when I teach listening, but it seems to be very time-consuming to gather things to use. How can I be more efficient in gathering the realia I need?

- For an introduction to the types of realia you can use at different levels, see James Abela's blog. He provides a helpful table for organising realia: http://www.jamesabela.co.uk/advanced/realia.html

- To explore creative uses of realia, see Simon Mumford's short essay on new uses of realia. This article takes a number of common objects – a bottle, a whistle, scissors, etc. – and shows how they can be used in grammar drills, listening and discussion activities: http://iteslj.org/Techniques/Mumford-Relia.html.

- For ideas on using realia in class, see the Busy Teacher blog. This entry describes nine ways in which teachers can use realia and gives five reasons for doing so: http://busyteacher.org/2842-realia-esl-classroom.html.

- For a similarly practical look at using realia in class as well as a brief description of the rationale behind it, see Jo Budden's blog entry: http://www.teachingenglish.org.uk/language-assistant/teaching-tips/realia.

Music

My students seem to enjoy it when I use music in the class, but I'm not convinced of the learning value. Why should I use music? How much should I use? How should I use music to teach?

- To investigate the theory behind music and language teaching, see Tim Murphey's classic resource book (Murphey 1992). Besides describing the links between music and language learning, Murphey provides numerous creative and enjoyable activities involving music and songs.

- Kevin Schoepp's essay describes affective, cognitive and linguistic reasons for using songs in the ESL classroom. See: http://iteslj.org/Articles/Schoepp-Songs.html.

- To explore uses of music in education generally, not only for language learning, see Chris Brewer's website, Life Sounds: www.musicandlearning.com. The site contains accessible articles investigating the effects of music for establishing a 'positive learning state', focusing concentration and improving memory.

Humour

I enjoy using jokes in my class, but sometimes I wonder if I'm just trying to entertain the students. Is humour a valuable tool for teaching?

- Humour can be used in L2 teaching for content presentations and to convey cultural and pragmatic information. For a study of the uses of humour in language teaching, see Wagner and Urios-Aparisi (2008).

- Humour can be a valuable tool for introducing and discussing cross-cultural prejudices. To explore this notion, see Ahmed Ahmed's frank short documentary on his life as an Egyptian-American comedian: http://www.youtube.com/watch?v=AexUlPFNcVw.

- Comic books and graphic novels often provide a window to understand culture. To hear a perspective on cross-cultural humour, listen to interviews in English with the Spanish comic book author Sergio Aragones. The discussion includes ideas related to cultural universals such as pantomime: http://www.youtube.com/watch?v=caHVIE6vltQ.

- Understanding humour has become an area of inquiry in sociolinguistics. Diane-Elaine Popa has researched the 'translatology' of humour. She talks about three components in understanding and translating humour: ability, situation and wordplay. You may be able to introduce these notions into your teaching. See http://www.dianapopa.net/DPopa%2002.pdf.

Top Down Frame

Pre-listening

Is pre-listening important? What kinds of pre-listening activities are most useful?

- It is now established that pre-listening activities can strongly affect comprehension. To explore how pre-listening variations tend to have differential effects, see Berne (1995). To explore **advance organisers**, see this Learn Net tutorial: http://www.projectlearnet.org/tutorials/advance_organizers.html and for a more specific introduction to **graphic organisers**, see John Handron's article http://www.glnd.k12.va.us/resources/graphicalorganizers/. For a more detailed discussion of **proactive interference**, see Lustig *et al.* (2001).

- **KWL charts** (what we **K**now, **W**ant to know, and **L**earned) can be used as pre-listening overviews of a listening passage and are likely, therefore, to improve attention and comprehension. See G.E. Tomkins's article at education.com: http://www.education.com/reference/article/K-W-L-charts-classroom/.

- Questions posed prior to listening exert an organising effect on comprehension and memory. Graesser and Person (1994) provide 14 categories of questions that teachers can ask to improve comprehension and critical thinking skills. As teachers, we may unwittingly focus on a few favourite types of questions but introducing a wider variety may engage the students more deeply.

- Studies of L2 listening comprehension consistently have shown the value of schema development prior to listening: building knowledge of key concepts helps students anticipate key information and has a positive impact on post-listening comprehension test scores (though with greater effect for advanced learners). See Chung (2002); Wilberschied and Berman (2004); Elkhafaifi (2005); Chang and Read (2007); Field (1998).

- Speculating on visuals related to the input prior to listening serves as an advance organiser and guides comprehension, as well as boosting post-listening recall. See Ginther (2002) and Herron et al. (1995). Question previews prior to listening also boost comprehension. See Elkhafaifi (2005) and Berne (1995).

Listening and memory

What can I do to help students develop their memory in the L2?

- Activities that require continuous processing involve the **phonological loop** in short-term memory. The best-known researcher in this field is probably Alan Baddeley, who describes a stage of short-term memory called the **visuo-spatial scratch pad** (or sketch pad), referring to the idea that information is stored and processed in a visual form. See Baddeley (2007) for ideas on how this form of memory develops.

- Increased **redundancy and repetition** is known to facilitate comprehension and memory, particularly if it is 'elaborative' redundancy. For more on the topic of elaboration and simplification and their effects on memory, see Jensen and Vinther (2003), Derwing (1996) and Rahimi (2011).

- Psychologist John Bransford, in discussing individual differences among learners, claims that some learners naturally tend to pay attention to overall **conceptual structure** as they listen or read, while others do not. He suggests that this individual difference in memory is important in predicting the relative success of learners in acquiring a second language. See Bransford (2003) and Bransford and Johnson (2004).

Listening and images

How does the use of visuals and visualisation affect listening?

- Visualisation is considered a type of cognitive listening strategy that allows the listener to process new information more efficiently. To explore the use of images for teaching purposes (and to obtain copyright-free visuals for use), see Keddie (2009); Wright (2003); Ruhe (1996); and Purdue University's website, Center for Technology-Enhanced Language Learning and Instruction http://tell.fll.purdue.edu/JapanProj/FLClipart/

- Speculating on visuals related to the input prior to listening serves as an advance organiser and guides comprehension, as well as boosting post-listening recall. See Ginther (2002) and Herron et al. (1995).

- Visualisation can be seen as a form of 'inner voice', a concept that has been discussed in relation to second language acquisition. Brian Tomlinson defines inner voice as 'a linguistic code we use to interact with sensory images and with affect in order to achieve a self-communication code.' See Tomlinson (2001).

Listening to the news

I know that the news is a great resource for learning listening. What are some of the best ways to exploit the news?

- Many teachers consider the news to be the best source of listening input – because there is *always* current news. To explore ideas for listening to the news, see Alan Mackenzie's tips: http://iteslj.org/Techniques/Mackenzie-CNN.html and David Bell's article on criteria for selection of news stories: http://www.cc.kyoto-su.ac.jp/information/tesl-ej/ej27/a2.html.

- Judith Frommer advocates using the news as a primary source for listening because it is always current and simulates 'real world listening challenges'. See Frommer (2006).

Academic listening and note-taking

What can I do to prepare my students for academic listening? Is note-taking helpful?

- Note-taking, particularly guided note-taking, is a widely used selective listening activity. For examples of note-taking in course materials that include appendices of note-taking strategies, see Clement *et al.* (2009), Kanaoka (2009) and Salehzadeh (2006).

- According to research on note-taking, only two **note-taking strategies** are consistently related to success, as measured by post-tests on content retention: (1) number of *content words* in notes and (2) *number of answers* to exam questions recorded in notes. To explore this finding, see Carrell *et al.* (2002) and Armbruster (2000).

- Although many college instructors provide students with online notes, it is not clear how access to these online notes affects student learning. Some results suggest that students receiving partial notes performed better on examinations than students receiving full notes. Students receiving full notes also self-reported more negative effects on attendance and motivation. To investigate this phenomenon, see Cornelius and Owen-DeSchryver (2008); Murphy and Cross (2002); and Neef *et al.* (2006).

Bottom Up Frame

Vocabulary and listening

Is listening a useful mode for learning vocabulary? How many exposures to a new word do students need in order to learn it? How can I help students recognise new vocabulary when they listen?

- One overriding principle for learning vocabulary is to increase the amount of engagement with lexical items. Such engagement includes encountering new vocabulary in a listening phase of an activity. For more on this topic, see Schmitt (2008) and Nation (2008).

- Gerry Luton presents an array of vocabulary techniques for use inside and outside the classroom, including online activities: http://www.cpr4esl.com/gerrys_vocab_teacher/teaching_suggestions.html

- Joseph Pettigrew presents a number of 'tips and techniques' for teaching vocabulary, most of which can be used as pre-listening **lexical priming** activities. See: http://people.bu.edu/jpettigr/Artilces_and_Presentations/Vocabulary.htm

- Vocabulary preparation (lexical priming) in particular is known to affect listening comprehension and confidence, as well as acquisition of new vocabulary. To explore the relative effects of different techniques of pre-viewing vocabulary, see Chang (2007).

Decoding connected speech

How can I help my students cope with connected speech - reductions, assimilations, elisions, allophonic variations?

- To explore issues of connected speech, it is useful to understand the differences between normal L1 listening rates (how many words per minute a typical L1 listener can understand) and rates at which L2 listeners can listen comfortably. Gary Buck provides a summary of research on this topic (Buck, 2001).

- Use of technology to enable learners to control the rate of speaking may improve their comprehension and their skills at recognising connected speech. One study has detailed how, when listeners can control the rate of the speaker through variable speed controls, their comprehension improves (Zhao, 2005).

- Although practice with 'fast speech' is necessary for L2 listening improvement, some researchers have found that, in an experimental setting, initial training with slow speech provides an advantage for training with connected speech because the listener builds up actual bottom up processing skills rather than fast speech decoding strategies. See McBride (2011) and Nakamura *et al.* (2008); Nishi and Kewley-Port (2007).

- Dyana Ellis developed an action research technique involving students listening to their own speech and identifying errors. This self-monitoring ability, she claims, leads to improved aural perception, generally. The improvements include a better perception of connected speech and an ability to recall and repeat progressively longer phrases. See Ellis (2008).

- Crawford and Ueyama provide a review of how reduced forms are covered in popular textbooks, including: listening to examples, repeating examples, listen and select (discrimination) exercises, listen and select citation versus reduced form, dictation, and mark forms that are reduced. See Crawford and Ueyama (2011); also see Brown and Kondo-Brown (2006).

- Identifying the segments in speech that learners have most trouble perceiving can be achieved through a process of **scaling**. This can be done through dictation tests in which you measure which syllables or words of a dictated sentence are most often perceived successfully. This can be a useful diagnostic technique to quantify mishearings, as you have clear evidence about which words and syllables are consistently misperceived. This type of diagnosis allows you to determine target forms for practice. See Rost (2011) for an example project.

- An increasing number of language teachers and teacher trainers are advocating a **noticing approach** to supplement listening instruction. The idea is that learners need to build up their perceptual abilities and not rely so much on top down processing. See Lynch (2006) and Wilson (2003).

Dictation and dictogloss

Is dictation a useful exercise for developing listening? What kinds of dictation are most effective?

- Dictation is a time-honoured teaching technique for integrating listening, grammar, vocabulary and writing. Because pure dictation of extended passages can be tedious and time-consuming, many teachers have developed variations to provide more efficient use of time, more interaction and clearer focus on listening. For a wealth of activities with imaginative twists on dictation, see Nation and Newton (2009, Chapter 4); Wilson (2008, Chapter 5); Brown (2011, Chapter 5), and Davies and Rinvolucri (1989).

- The **dictogloss** method, which is comprised of student reconstruction of content-rich input, is seen as part of a processing approach to language instruction. For more on this topic, see: VanPatten *et al.* (2009), Prince (2012) and Vasiljevic (2010). You may also wish to consult a seminal work in this area by Ruth Wajnryb (1990).

Testing listening

What are some effective ways to test listening? How can I help my students perform well on standardised listening tests?

- Many listening tasks can be used for assessment if you include some kind of scoring rubric. See Josh Kurzweil's article on oral quizzing: http://iteslj.org/Techniques/Kurzweil-OralQuizzing.html. Kurzweil describes the positive **washback effect** that this kind of quiz can create.

- Interactive tasks can be recorded for assessment of interactive listening skills. See ESLgo's website for Jim Trotta's ideas about creating a speaking scale for oral assessments. A lot of Trotta's ideas come from working with groups of university students in Korea: http://www.eslgo.com/resources/sa/oral_evaluation.html

- It is useful for teachers to be aware of the structure and content of any standardised listening tests that their students will be taking. You can use practice tests in your teaching from time to time to research students' progress.

  First Certificate English (FCE):
  http://www.examenglish.com/FCE/fcelistening.htm

  International English Language Testing System (IELTS):
  http://www.examenglish.com/IELTS/IELTS_listening.htm

  Test of English as a Foreign Language (TOEFL):
  http://www.examenglish.com/TOEFL/toefl_listening.htm

  STEP (Standardized Test of English Proficiency):
  http://www.elllo.org/english/STeP.htm

Interactive Frame

Storytelling

I'd like to do more storytelling in my classes, instead of using recordings. What storytelling techniques are best?

- Penny Ur was one of the first listening experts to advocate maximum use of teacher 'live input' in listening classes. She called this 'real life listening'. She provides many tips for storytelling. See Ur's classic work, still in print (Ur, 1984).

- Short anecdotes take little time to prepare. To explore the use of anecdotes in language classes, see a webinar by Sue Kay and Vaughan Jones (web-based seminar): http://www.youtube.com/watch?v=CRnBC7F_WB0.

- To explore the use of stories in EFL, see Odilea Rocha Erkaya's essay on the benefits of using short literary extracts in class (using the four skills, increasing student motivation, developing higher-order thinking skills and examining aspects of culture): http://www.asian-efl-journal.com/pta_nov_ore.pdf.

- For a collection of suggestions of genres and enjoyable activities using stories for the language classroom, see Wajnryb (2003). The book contains different genres – soap operas, news stories, myths and legends, etc. – and gives clear, straightforward recipes on how to use them in class.

- Mario Rinvolucri calls storytelling the language teacher's oldest technique. To read Rinvolucri's ideas about adapting stories for use in language teaching, see: 'Adapting Stories for Language Teaching' at http://www.teachingenglish.org.uk/articles/story-telling-language-teachers-oldest-technique.

- Even teachers with no drama training can learn to tell engaging stories. To explore factors in telling a compelling story, see Chris King's article, 'Story Telling Power: What makes a good story?' at http://www.creativekeys.net/storytellingpower/article1004.html.

- Brian Sturm talks about 'story trance' and why stories are a powerful vehicle for teaching. To compare the power of stories across cultures, see: 'The Power of Stories' at http://www.youtube.com/watch?v=UFC-URW6wkU&feature=related.

- Many teachers like to mine their own personal stories as sources of listening material. To explore teaching ideas that use personal narratives, see Griffiths and Keohane (2000).

Listening and interaction

Is interaction important for developing listening? Some students don't get along with other students in class and some seem excessively shy or disconnected – what can I do to address these issues?

- Most learners need instruction and continuing encouragement to collaborate actively during tasks. Nation and Newton (2009) provide a detailed set of guidelines for getting reluctant students to speak in interactive tasks. They address the main causes of reluctance and provide practical solutions.

- Identity is a key factor influencing interaction. Communication in an L2 is more complex than communication in an L1. L2 communication involves all of the issues of L1 communication, such as building relationships, creating a positive climate for interaction and self-disclosure. It also involves issues of identity as an L2 speaker-listener and 'investment' in learning a new communication system. Many language educators advocate exploring issues of identity as they help students develop their interactive abilities. The two (exploring identity and improving interactive ability, tend to go hand-in-hand. See Adler *et al.* (2004), Norton (2010) and Morgan and Clarke (2011).

- Interaction can be understood as part of a **collaborative learning** approach. To explore collaborative learning, see Nunan (1992) and McCafferty and Iddings (2006). Collaborative learning continuously demands interaction, negotiation of meaning and pushed output, all of which are claimed to promote language acquisition.

- There may be motivational effects to including interaction as part of a language class. Interactive tasks may help classes develop 'emotional intersubjectivity' and push the entire group to do more 'knowledge **co-construction**' and proceed more effectively towards their language goals. See Imai (2010) and Jacobs and McCafferty (2006).

- Development of interactive skills can also lead to more harmonious and more equitable communication, in addition to the language acquisition advantages that accrue. Dennis Rivers has developed a successful approach to transforming ineffective conversations by confronting seven challenges to communication, three of which directly entail attitudes about listening. See: http://www.newconversations.net/sevenchallenges.pdf.

Supportive communication practices – specifically the intent to contribute to our conversation partner's well-being – influence our listening performance, even in difficult circumstances. For insight into this issue, see the interview with communication specialist Marshall Rosenberg: http://www.youtube.com/watch?v=-dpk5Z7GIFs.

- Many second language acquisition researchers recommend that listening instruction include a wide range of oral interaction tasks that present a need and opportunity not only for negotiation of meaning but also pushed output. 'Pushed' output refers to requiring learners to articulate complex ideas that normally they may opt to express non-verbally. For more on this topic, see Lynch (2009), Maleki (2007), Pica (2005) and Batstone (2002).

Listening and responding

How important is responding in listening? Is speed of response important? How do L2 listeners get more control of a conversation with a native speaker?

- In a symmetrical conversation, both parties have equal rights and equal means for gaining the floor and directing the discourse. Most L2 learners struggle with knowing how and when to intervene in a conversation and there are often social and even political pressures preventing them from actually intervening. To explore contextual strategies for L2 learners, see Block (2003) and Bremer *et al.* (1996).

- Murphey reports on a communicative methodology he built around shadowing techniques, which he has used widely in Japan, Korea and China. See Murphey (2000).

- Shadowing is a form of backchannelling, that is, providing the speaker cues as to how the message is being received by the listener. To explore this topic further, see LoCastro (1987) and Fujimoto (2009).

- Reidsma *et al.* (2010) show how computer simulations of human interaction include shadowing. Even in computer-mediated communication, shadowing gives the user the experience of human interaction.

- Learning to backchannel can create an immediate increase in perceptions of fluency in L2 listeners. For an experimental study, see Wolf (2008).

- Using controversial topics presents an opportunity for working with the **pragmatics** of disagreeing. See Bardovi-Harlig (2006).

- Some evidence supports claims about the value of teaching **formulaic English** for students in conversation partner programmes. When using colloquial formulae, speakers can be confident that the speech act performed will be understood. In addition, the speakers are often perceived by native speakers as more confident when they use such formulaic language. See Bardovi-Harlig (2010).

Autonomous Frame

Using the web and mobile devices for listening

What are the best uses of the internet for improving learners' listening? I'd like to encourage students to use their mobile devices to improve their listening. What can I do?

- One way to exploit internet resources is through task-based interaction. Yilmaz and Granena examined the potential of learner–learner interaction through Computer-Mediated Communication (CMC) to focus learners' attention on form. They discovered that the use of jigsaw tasks and collaborative tasks promoted gains in accuracy more than open communication activities. See Yilmaz and Granena (2010).

- The internet offers opportunities for communicative activities that have similar, if not equal, value to live communicative activities. Choe (2011) investigated the effects of the cognitive complexity of tasks and pair proficiency on task production in CMC, with some surprising results.

- Webquests are organised research 'missions' that students conduct on the web, usually with teachers pre-approving the websites that will be accessed. If webquests include research involving listening and viewing, the students will be developing their listening ability. For an introduction to setting up **webquests**, see http://webquest.org/; http://webquest.sdsu.edu/about_webquests.html and http://zunal.com/.

- The prevalent use of mobile devices is changing the way that people view learning, including language learning. To explore this area, see Rogers and Price (2009). You can also search for downloadable listening apps, including many that are free at iTunes: http://itunes.apple.com/us/app/ and one that we designed: 'Listening Master': http://itunes.apple.com/us/app/listening-master/id505756112?mt=8.

Listening strategies

What are listening strategies? Is it useful to teach them?

- In the context of second language learning, one definition of listening strategies is a skill that a competent first language listener possesses and uses automatically, but which a second language listener resorts to consciously to compensate for incomplete ability. Another definition is *any conscious plan that a learner uses to improve* his or her comprehension or listening performance. If we go along with this second definition, then we can state clearly that *it is essential to coach students to make conscious plans for how to improve their listening!* It is not necessary to call this strategy development or strategy coaching, but an important part of listening instruction is including specific plans for improvement that students can try out during an activity.

  There are different ways of categorising listening strategies. One common categorisaton scheme uses three general categories, with several exemplars of each strategy: **cognitive** (predicting/inferencing, elaboration, contextualisation, imagery, summarisation, translation, deduction, fixation); **metacognitive** (planning, comprehension monitoring, directed attention, selective attention, evaluation); **socioaffective** (questioning, cooperation, anxiety reduction, relaxation). While this scheme may be useful for understanding strategy development, we have found that, for instructional purposes, it is useful to frame listening strategies into categories that students can most readily understand. We use eight basic strategies, with several exemplars of each one: planning, focusing attention, monitoring, evaluating, inferencing, elaborating, collaborating and reviewing. See Appendix 1 for the complete list. If you identify one or two strategies that students can keep in mind for each listening task, it will be easier for students to try them out and evaluate their utility.

- Second language acquisition research has shown that use of **communication strategies** allows 'early communication', that is, even beginners can begin to use authentic conversation for learning purposes, if they have strategic tools. See Weinart (1995) and Krashen *et al.* (1984).

Listening in the community

*My students **live** in an English-speaking community, but don't take advantage of the resources for listening outside of class. What can I do?*

*My students **don't live** in an English-speaking community, but there are a lot of English speakers that they can access if they wish. How can I take advantage of the opportunities that are out there?*

- There is a developing trend for students to interact with the community as part of both their personal and also their language development. Some schools are now incorporating service learning elements in the curriculum. See Natalie Russell's article on connecting learners with the community (Russell, 2007) for a description of a successful project with adult ESL students interacting with their communities. Moser and Rogers (2005) describe a similar initiative involving high-school students.

- Purmensky (2009) outlines an approach to second language teaching based on service learning. She defines service learning as a form of instruction that involves mutual benefit for both the learner and the community, and provides a sample syllabus.

- Camden College of English in London devised an award-winning course called English Language Cultural Experience that involves the students going on visits every day to different venues, such as museums and galleries, and interacting with the public. See: http://www.camdencollege.com/cultural_experience_london.html.

Extensive listening

I've heard about extensive listening, but I don't think my students are ready for listening on their own outside of class. How can I start?

- Rendaya and Farrell (2011) conducted a study with low-intermediate students, documenting the self-reported effects (mainly positive) of adding extensive listening to the curriculum. Their article provides a useful programme outline plus a series of questionnaires.

- Rob Waring, one of the original proponents of extensive listening, believes an extensive listening programme can be effective from intermediate level. He provides background and resources for guiding students into an extensive listening programme, with helpful troubleshooting tips for keeping the students focused: http://www.robwaring.org/el/.

- Linking reading with listening is one way to introduce extensive listening. This linked approach has been studied, with reports of beneficial learning and motivation effects. Woodall (2010) reports significant gains in vocabulary acquisition for groups with reading-while-listening supplementary instruction. For more on this topic, see Brown (2011); Brown *et al.* (2008) and Chang (2009).

- Task design is an important aspect of a successful extensive listening programme. Providing selective listening tasks is a technique to help L2 learners increase their attention span for the whole length of the text. See Goh (2000) and Hasan (2000). Multiple listenings with associated tasks is a similarly effective approach for boosting comprehension and motivation to continue, as each listening serves as a pre-listening for the subsequent attempt. See Iimura (2007) and Sakai (2009).

- The **sheltered instruction** approach involves extensive listening outside of class, along with interaction and application steps. See Rost (2011: 193–197) for steps in creating and adapting this type of approach.

Subtitles and transcripts

Should I use subtitles and transcripts with my students?
If so, how and when?

- Susan Gass and her colleagues have studied the use of subtitles in foreign language learning for several years. They contend that subtitles and transcripts are powerful teaching tools when used properly. See Winke *et al.* (2010) for a review.

- To explore the practical use of transcripts, see Wilson (2008). He gives a brief rationale for using transcripts and provides some simple activities for their exploitation.

- There are varying views about the use of captioning and help menus (e.g. vocabulary glosses) when teaching with multimedia and multiple modes of presentation. One view is that the additional support will increase context, reduce cognitive load and improve comprehension. See Grgurović and Hegelheimer (2007); Clark *et al.* (2006); and Jones and Plass (2002).

Presentations and guest speakers

How can my students get the most out of presentations, whether they're by guest speakers or by their classmates?

- For tips on getting the most out of your guest speakers, see: http://busyteacher.org/7083-top-10-ways-get-most-from-guest-speaker.html. The tips include ideas for preparation of the students and the speaker, debriefing the students and ways of maintaining ongoing relationships with the guest speaker.

- The sociolinguistic phenomenon of '**accommodation**' describes how speakers and listeners 'normalise' their speech so that their interlocutors can better understand them. To explore how this phenomenon affects classroom teaching, see: Ross and Berwick (1992) and Rogerson-Revell (2010).

- For more about giving effective presentations, and promoting fuller audience participation, see Christine Bauer-Ramazani's resource page: http://academics.smcvt.edu/cbauer-ramazani/cbr/iep/spkg.htm. Her framework focuses on the energy and interactiveness of the speaker.

- To learn about the uses of the Pecha-Kucha format for presentations, visit Vernon Melbourne's teaching website: http://teachesltech.vfowler.com/2010/12/pecha-kucha-presenting/.

Glossary

accommodation When two parties in an interaction compromise towards the communication norms of the other (including speed of speaking, vocabulary use, body language, etc.).

action research A type of research in which the researchers (in this case, teachers) are participants, examining the effects of their own practice on the students' performance in class. The goal is to use a cycle of observation and experimentation to implement change.

activation spaces Neural patterns that are activated in response to certain words and ideas.

active learner hypothesis The idea that great attention, effort and persistence on the part of the learner towards achieving goals will lead to a higher level of motivation and performance in language learning.

active listening A range of cognitive, emotional and physical activity that places the listener in the position of engaged participant in the listening process.

advance organiser An instructional tool that presents background information (or the rhetorical structure of a passage) graphically and symbolically, thereby helping students to anticipate what they will hear. Advance organisers may come in different forms including charts, diagrams, tables and lists.

affective filter A proposed part of the sensory processing system that subconsciously screens incoming language for factors related to the perceiver's feelings, attitude, motives and needs.

Affective Frame A category of learning that examines success in listening in terms of motivation and personal engagement. It involves a focus on the emotional states of listeners and their psychological readiness to listen successfully.

applied research The systematic collection of data used by practitioners for a specific, practical purpose.

appropriation The notion of students personalising and taking ownership of new language, ideas or texts, by manipulating them for their own ends or for a classroom task, e.g. doing further research or presenting on the topic. Appropriation, by allowing deeper processing of new language, aids acquisition.

assimilated Refers to the way sounds become modified when they are articulated in succession, often resulting in merged word boundaries. For example, *right back* in connected speech sounds like *ripe back*.

automatic word recognition Recognising words without needing to pause and consciously attempt to retrieve the meaning from memory.

Autonomous Frame A learning category that examines success in listening in terms of locating and using listening resources outside the classroom. It involves a focus on independent learning and the development of strategies that allow students to deal with the challenges of authentic listening input.

backchannelling Short verbal and non-verbal messages sent by the listener during or slightly after a partner's speaking turn. Backchannelling is used to indicate the listener's attention to the speaker.

Bottom Up Frame A learning category that examines success in listening in terms of the accurate perception of sounds and words and the ability to build the message using these speech signals. It involves a focus on decoding the speech signal as it occurs in real time.

co-construction Working together to build ideas or meaning.

cognitive strategies Techniques used by students to improve their performance in a language learning task. These might include brainstorming key vocabulary before listening, guessing unknown words from the context, monitoring comprehension, writing notes while listening, etc.

collaborative learning When two or more students work together to learn something, pooling their different skills and resources. Collaborative learning involves interaction, negotiation of meaning and group responsibility for learning.

collaborative strategies Techniques that involve students working together in order to achieve a task. In a conversation, these techniques might include repetition, slowing down, asking for confirmation, etc.

communication strategies Techniques that are used to overcome problems in communication. These might include asking for clarification, paraphrasing, switching to L1, etc.

conceptual structure The way in which a text or information is organised. e.g. if the information is presented in chronological order, hierarchical order, etc.

dictogloss A learning activity in which students listen for an extended period and then attempt to reconstruct what they heard, and later compare their version with the original.

elided Refers to the omission of sounds in rapid connected speech. This is usually the result of one word 'sliding' into another, and the sound omitted is usually an initial or final sound in a word. An example: the '*t*' in *next please* is elided so that the expression sounds like *necks please*.

extensive listening A form of listening practice in which the input material is considerably longer than most classroom texts. The students may engage in extensive listening for pleasure or they may perform comprehension-oriented tasks.

fast speech phenomena Features of rapid connected speech (see *assimilated* and *elided*). These typically involve words becoming distorted, with sounds omitted or changed due to the speed at which they are uttered.

fillers Sounds and words that are used to fill silences during a speaking turn. These sounds and words carry no specific semantic meaning.

formal schema The way in which the information in a piece of discourse is typically organised. For example, a text may be organised in a 'cause-effect' or a 'problem-solution' structure. The organisation is signalled through rhetorical devices that the listener needs to perceive in order to listen successfully.

formulaic English Sequences of words that are interpreted as having a single meaning, e.g. idioms, collocations, set phrases and proverbs. These prefabricated chunks are stored and retrieved in the memory as whole units.

grammar noticing approach An approach to teaching that involves the students paying conscious attention to a feature of grammar that occurs (usually repeatedly) in a text or passage.

graphic organisers Instructional tools that are often used to activate the students' expectations about a topic. Graphic organisers help students to anticipate the content of a text or passage.

guided note-taking An exercise that uses incomplete notes prepared by the teacher. While listening, the students complete the notes with key concepts, definitions and other information. Besides words, the cues in the guided notes might include bullet points, numbered lists, charts and concept maps.

inferential processes Using background knowledge and reasoning to complete missing or obscure parts of a text or passage.

integrated skills approach An organisational approach to a language curriculum that involves teaching 'the four skills' (reading, writing, speaking and listening) in conjunction with one another rather than focusing on only one skill.

interaction hypothesis The idea that face-to-face communication is central to language acquisition because it allows for learners to create comprehensible input.

Interactive Frame A learning category that examines success in listening in terms of collaborating with speakers and providing meaningful feedback. It involves a focus on co-construction of meaning and interdependence between speaker and listener as a requisite of successful listening.

interactive transcript A script, of a recorded dialogue, for example, that is displayed alongside the video or audio source, similar to subtitles, and allows the viewer to navigate to parts of the video or audio by clicking on the text.

interpersonal processes Working with others to achieve tasks and to learn. The processes typically involve collaborative activities and negotiation of meaning.

intonation patterns Features of speech involving modification of the speaker's pitch to express meaning or functionality.

KWL charts A table with three columns that can be used to organise students' prior knowledge about a topic before they listen. The columns are named K, W and L. K stands for **K**now (what the students know about the topic); W stands for **W**ant to know (what the students want to know about the topic); and L stands for **L**earned (what the students learned about the topic from listening).

learning styles The varying ways in which different people learn best. These include an auditory learning style (people learn best by listening), a visual learning style (people learn best by looking), an interpersonal learning style (people learn best working with others), etc.

lexical priming Previewing items of vocabulary that will occur in a listening passage. Lexical priming, by alerting students to key words and phrases, helps them to comprehend the passage.

mediation In language assessment, this refers to the translation of ideas from the learner's L1 to L2 or vice-versa.

metacognitive strategies Plans or approaches used consciously by students to improve their learning opportunities and overall success. In a general language learning context, these include keeping a vocabulary notebook, using a dictionary, watching movies in the target language, etc. Examples of metacognitive listening strategies are comprehension monitoring, selective attention and evaluation of one's success in a comprehension task.

motivational processes The development of positive attitudes towards task-achievement and learning. These may be stimulated by the student's mood, by teacher action, by interesting tasks, or a variety of other factors.

neural pathways Connections between different parts of the brain. Neural pathways are constructed as we learn new concepts and ideas.

note-taking strategies Techniques for taking notes during a lecture to improve comprehension and memory.

noticing approach A teaching approach that encourages students to pay attention to a feature of language that occurs in a listening passage.

paralinguistic signals Sound modulations made by speakers involving stress, loudness, intonation, breathing patterns, style of articulation, and tone of voice. These affect the meaning of what is said, e.g. showing the speaker's attitude or the intended importance of the message.

Pecha-Kucha A format for giving presentations, involving 20 slides shown for exactly 20 seconds each. The speaker delivers the presentation using the slides as guidance or illustration, and attempts to continue talking to keep up with the speed of the slides (*pecha-kucha hanasu* is a Japanese expression meaning 'talk constantly').

perceptual processes Meaning-oriented responses to sensory stimulation, e.g. sound waves. On perceiving sound, the listener automatically begins to make sense of it, matching the stimulus with prototypical sounds and combinations of sounds that constitute known words.

phoneme The basic element of spoken language, from which words are constructed.

phonological loop A memory process that helps us to deal with auditory information. The phonological loop allows us to hold sounds in our memories for a few seconds after they have been uttered. This helps us to recognise words and phrases.

phonotactic system The organisation of language that dictates which phonemes can and cannot be combined. In all languages, there are different clusters of consonants and vowels that cannot occur.

pragmatics The study of speech acts in real contexts. The field is concerned mainly with the context of an utterance, e.g. the situation, and the speaker's mood, intentions and relations with the listener.

proactive interference A situation in which the learning of 'new information' is made more difficult because of the presence of 'old information' stored in the memory.

prototypes Cognitive models of concepts. These consist of images or ideas that we typically associate with a given situation or subject.

pushed output Complex ideas that students are required to articulate, often using structures they have not yet fully acquired.

realia Real objects that are brought into the classroom and used for various pedagogical reasons, e.g. to illustrate vocabulary or provide stimulus for a task.

redundancy and repetition Features of authentic speech involving re-using the same words, phrases and concepts to improve listener comprehension. The use of redundancy and repetition allows listeners more processing time.

reflective listening A style of listening designed to help a speaker deal with a charged emotional state. After hearing the speaker's message, the listener 'reflects' the message (by paraphrasing) in order to invite recognition, confirmation, clarification or elaboration.

reliability A feature of tests, when they give consistent results, regardless of extraneous factors such as the identity and mood of the person grading them, or environmental factors.

repair A type of strategy used to rectify difficulties in communication. For example, a speaker may offer an alternative phrase or attempt to explain what was meant.

scaling Measuring perception difficulties of a group of learners by mapping which parts of an input (e.g. in a dictation) are most often perceived correctly or incorrectly. This type of diagnosis allows teachers to isolate target forms for practice.

schema activation Stimulation of the underlying organisational patterns or conceptual frameworks by which we perceive the world. The schema is a mental structure of preconceived ideas or a representation of our knowledge and assumptions about something.

selective listening A form of listening practice in which students listen only for key information or for specific pre-selected items.

self-management strategies Techniques for organising and planning your own learning. These include setting up personal schedules, monitoring individual improvement, evaluating courses of action, etc.

service learning A process in which instruction and reflection is integrated with community service in order to enrich the learning experience. Service learning often involves practical, real-life application of things learned in the classroom.

shadowing Repeating all or part of what the speaker has said. Shadowing is a technique that allows students to notice phonological features of the target language, as well as encouraging close listening.

sheltered instruction An approach to teaching English language learners that integrates content and language instruction. Sheltered instruction is often used in the USA's elementary, middle and high schools as a way to develop language skills while simultaneously allowing students to learn appropriate academic content.

socioaffective strategies Techniques that help to improve motivation and performance in language learning, such as working with others, asking questions, cooperating and developing positive attitudes.

spaced repetitions The occurrence of the same word at regular intervals (in a listening passage). Spaced repetitions allow students to acquire vocabulary with greater ease.

speech rate The number of words spoken per minute. A normal speech rate of a fluent speaker with familiar content is about 150 words per minute.

strategic processes Courses of action (see *cognitive, metacognitive* and *socioaffective strategies*) that are planned by learners in order to improve performance.

supportive communication A conversational style, the main goal of which is to contribute to a partner's well-being. This may involve avoiding judgment, showing interest through body language, encouraging free expression, asking for elaboration, etc.

think aloud protocols A method used to gather data. Listeners are asked to 'think aloud', describing what they are doing, feeling or thinking while performing a set task. Think aloud protocols may shed light on how students go about understanding texts or listening passages.

Top Down Frame A learning category that examines success in listening in terms of using background knowledge of the topic or context to anticipate and interpret the input. It involves a focus on the listener's expectations.

top down processing A way of considering and interpreting listening input based on our life experiences and background knowledge of the topic. We use our expectations and experiences to construct mental representations of the input.

validity The degree to which a process or outcome – particularly regarding testing procedures – is justifiable, effective, logical and fair.

visuo-spatial scratch pad A short-term memory process that allows for the creation, manipulation and temporary storage of visual and spatial information, such as the shapes, sizes and colours of objects.

washback effect The consequences of an action on subsequent attitudes of participants (e.g. students). In educational settings, this involves the tendency of teaching goals to mirror testing goals.

webquests An instructional activity involving students searching through web material in order to fulfill a task. The teacher sets a higher-order question or task and the students use pre-selected online resources to answer it.

Appendix 1
Active listening strategies

Strategies are conscious plans that the learner applies to improve his or her performance. Active listening strategies are those strategies that the learners intentionally use to improve comprehension, increase their ability to respond to the speaker or use information the speaker has provided, or complete a listening task or test successfully. The list here provides strategies in eight categories, all of which are practised in the listening activities in *Active Listening*. The strategy targets are provided in the 'Aims' section of each Activity.

(Based on Rost, 2011, 1990; Salehzadeh, 2006; Vandergrift and Goh, 2012; Vandergrift, 1997; O'Malley and Chamot, 1990; Oxford, 2011)

1. Planning
Developing an awareness of the steps needed to accomplish a listening task, anticipating content that may be introduced, coming up with an 'action plan'

Advance organising: Clarifying the goals of a task before listening

Self-management: Rehearsing the steps to take to deal with a listening task

2. Focusing attention
Concentrating on the input and task at hand, avoiding distractions and disruptive thoughts, reminding oneself of the plan

Directed attention: Attending to the listening task, consciously ignoring distractions or any tendency to give up

Selective attention: Attending to specific aspects of the listening input, such as key words or ideas that have been anticipated

Persistent attention: Attending to broad meaning, keeping flow of attention even if temporarily distracted by unknown language

Noticing attention: Attending to new language, specific language, rhetorical forms in the input

3. Monitoring
Verifying or adjusting one's understanding or way of understanding during a task

Comprehension monitoring: Checking how well one is understanding, identifying problematic aspects in the input

Double-check monitoring: Verifying one's initial understanding and making revisions in understanding as needed, during the second listening to the same input

Emotional monitoring: Keeping track of one's feelings, encouraging oneself to keep listening, finding ways to counter negative emotions and anxiety

4. Evaluating
Checking the outcome of one's listening process against a standard of accuracy or completeness

Performance evaluation: Checking one's overall attainment of the task goals

Problem evaluation: Identifying what specific issue needs to be solved or understood or what part of the listening task still needs to be completed

Revision evaluation: Choosing a second listening to assist understanding or selecting an alternative way of accomplishing a listening task

5. Inferencing
Using information in the input to guess the meaning of unfamiliar language, to predict content, or to fill in missing information

Linguistic inferencing: Using words you know to guess meanings of unknown words or blurs of sounds

Contextual inferencing: Consciously using knowledge of the setting and extralinguistic features (items in the environment) to create or amplify meaning

Speaker inferencing: Using the speaker's tone of voice, paralinguistic cues (stress, pause, intonation) or to guess intended meanings, facial expressions, body language and baton signals (hand and arm movements) to guess intended meanings

Multimodal inferencing: Using background sounds, visual cues, supplementary text and intuitive sense of the input to infer meaning

Predictive inferencing: Anticipating details in a specific part of the input (local prediction), or anticipating the gist of what is coming in the input (global prediction)

Retrospective inferencing: Thinking back over a large chunk of input to fill in gaps and consolidate one's understanding

6. Elaborating
Using prior knowledge from outside the input and relating it to content in the input in order to enrich one's interpretation of the input

Personal elaboration: Connecting with prior personal experiences

World elaboration: Connecting with knowledge gained about the world

Creative elaboration: Making up background information to contextualise the inputs, generating questions that relate to the input, or introducing new possibilities to extend the input

Visual elaboration: Using mental visualisations to represent aspects of the input

7. Collaborating
Cooperating with the speaker, other listeners, and outside sources for assistance with improving comprehension or enriching interpretations

Seeking clarification: Asking for repetition, explanations, rephrasings or elaborations about the language just heard

Seeking confirmation: Asking for verification that what you have said has been understood

Backchannelling: Showing the speaker that you are engaged, following the input and ready to continue listening

Joint task construction: Working together, with the speaker or with another listener to solve a problem or complete a task

Resourcing: Using available referencing resources to deepen language and idea comprehension

8. Reviewing
Condensing, reordering or transferring from one modality to another, of what one has processed to help understanding, memory storage and retrieval

Summarisation: Making a mental or verbal (oral or written) summary of information presentation

Repetition: Repeating or paraphrasing part of what was heard as part of a listening task

Noting: Writing down key words or ideas in another form (abbreviated verbal or graphic form) to assist in recall or performance of a task

Mediating: Rendering ideas from the input to the listener's L1, orally or in writing

Appendix 2
Audio files to accompany activities

Please visit www.routledge.com/9781408296851 for audio recordings of these transcripts:

Audio file 1: Photo Album

Audio file 2: Emotional Scenes

Audio file 3: Creative Visualisation

Audio file 4: Finish the Story

Audio file 5: Memories

Audio file 6: 2-20-2 Pictures

Audio file 7: The Right Thing

Audio file 8: Split Notes

Audio file 9: Race to the Wall

Audio file 10: Action Skits

Audio file 11: Pause and Predict

Audio file 12: Whisper Dictation

Audio file 13: Interrupted Story

Audio file 14: My Turn/Your Turn

References

Adler, R., Rosenfeld, L. and Proctor, R. (2004) *Interplay: The Process of Interpersonal Communication*. New York: Oxford University Press.

Ahmed, T. (2009) 'Language learning motivation: What's on a student's mind when learning English language?' *BRAC University Journal*, 6, pp. 81–92.

Alptekin, C. (2002) 'Towards intercultural communicative competence in ELT', *ELT Journal*, 56, pp. 57–64.

Altenberg, E. (2005) 'The perception of word boundaries in a second language', *Second Language Research*, 21, pp. 325–58.

Aniero, S. (1990) 'The influence of receiver apprehension among puerto rican college students', PhD thesis, New York University, in *Dissertation Abstracts International*, 50, 2300A.

Anstey, M. and Bull, G. (2006) *Teaching and Learning Multiliteracies: Changing Times, Changing Literacies*. Washington, DC: International Reading Association.

Aragao, R. (2011) 'Beliefs and emotions in foreign language learning', *System*, 39, pp. 302–313.

Armbruster, B.B. (2000) 'Taking notes from lectures' in Flippo, R. and Caverly, D. (eds), *Handbook of College Reading and Study Strategy Research*, pp. 175–199. Mahwah, NJ: Erlbaum.

Asher, J.J. (2003) *Learning Another Language Through Actions*. Sixth edition. Los Gatos, CA: Sky Oaks Production.

Asher, J.J. (1969) 'The Total Physical Response approach to second language learning', *Modern Language Journal*, 53(1), pp. 3–17.

Asher, J.J. (1966) 'The learning strategy of the Total Physical Response: a review', *Modern Language Journal*, 50(1), pp. 79–84.

Bacon, S.M. (1992) 'The relationship between gender, comprehension, processing strategies, and cognitive and affective response in foreign language listening', *Modern Language Journal*, 76, pp. 160–178.

Baddeley, A.D. (2007) *Working Memory, Thought and Action*. Oxford: Oxford University Press.

Bardovi-Harlig, K. (2006) 'On the role of formulas in the acquisition of L2 pragmatics' in Bardovi-Harlig, K., Felix-Breasdefer, J. and Omar, A. (eds) *Pragmatics and Language Learning*, Volume 11. Honolulu: NFLRC.

Bardovi-Harlig, K. (2010) 'Recognition of conventional expressions in L2 pragmatics' in Kasper, G. Nguyen, H. and Yoshimi, D. (eds) *Pragmatics and Language Learning*, Volume 12. Honolulu: NFLRC.

Bar-On, R. (1997) *Bar-On Emotional Quotient Inventory: User's Manual*. Toronto: Multi-Health Systems.

Barrs, K. (2012) 'Fostering computer-mediated L2 interaction beyond the classroom', *Language Learning and Technology*, 16, pp. 10–25.

Bartlett, F.C. (1932) *Remembering: A Study in Experimental and Social Psychology*. Cambridge: Cambridge University Press.

Batstone, R. (2002) 'Contexts of engagement: a discourse perspective on "intake" and "pushed output",' *System*, 30, pp. 1–14.

Bell, N. (2009) 'Learning about and through humor in the second language classroom', *Language Teaching Research*, 13, pp. 241–258.

Benet-Martínez, V. and Lee, F. (2009) 'Exploring the socio-cognitive consequences of biculturalism: Cognitive complexity' in Gary, A. and Milonas, K. (eds) *From Herodotus' Ethnographic Journeys to Cross-cultural Research*. Athens: Atrapos Editions.

Benson, P. and Chik, A. (2010) 'New literacies and autonomy in foreign language learning' in Luzón, M., Ruiz-Madrid, M. and Villanueva, M. (eds) *Digital Genres, New Literacies and Autonomy in Language Learning*, pp. 63–80. Newcastle: Cambridge Scholars Publishing.

Berne, J. (1995) 'How does varying pre-listening activities affect second language listening comprehension?' *Hispania*, 78, pp. 316–329.

Berne, J. (2004) 'Listening comprehension strategies: a review of the literature', *Foreign Language Annals*, 37, pp. 521–533.

Bialystok, E. and Craik, F. (2010) 'Cognitive and linguistic processing in the bilingual mind', *Current Directions in Psychological Science*, 19, pp. 19–23.

Biber, D. (2009) 'A corpus-driven approach to formulaic language in English: multi-word patterns in speech and writing', *International Journal of Corpus Linguistics*, 14, pp. 275–311.

Binder, J., Desai, R. and Graves, W. (2009) 'Where is the semantic system? A critical review and meta-analysis of 120 functional neuroimaging studies', *Cerebral Cortex*, 18, pp. 2767–2796.

Block, D. (2003) *The Social Turn in Second Language Acquisition*. Edinburgh: Edinburgh University Press.

Boal, A. (1979) *Theatre of the Oppressed (Teatro de Oprimidio)*. New York: Theatre Communications Group.

Bolitho, R., Carter, R., Hughes, R., Ivanic, R., Masuhara, H. and Tomlinson, B. (2003) 'Ten questions about language awareness', *ELT Journal*, 57, pp. 251–60.

Boroditsky, L. (2001) 'Does language shape thought? Mandarin and English speakers' conception of time', *Cognitive Psychology*, 45, pp. 1–22.

Bowe, H. and Martin, K. (2007) *Communication Across Cultures: Mutual Understanding in a Global World*. Cambridge: Cambridge University Press.

Bransford, J. (2003) *How People Learn: Brain, Mind, Experience, and School*. Washington, DC: National Academies.

Bransford, J. and Johnson, M. (2004) 'Contextual prerequisites for under-standing: some investigations of comprehension and recall' in Balota, D. and Marsh, E. (eds) *Cognitive Psychology: Key Readings*. New York: Psychology Press.

Brazil, D. (1995) *A Grammar of Speech*. Oxford: Oxford University Press.

Bremer, K., Roberts, C., Vasseur, M., Simonot, M. and Broeder, P. (1996) *Achieving Understanding*. London: Longman.

Briggs Myers, I. (1980) *Gifts Differing: Understanding Personality Type*. Mountain View CA: Davies-Black Publishing.

Brill, J. and Park, Y. (2008) 'Facilitating engaged learning in the interaction age: taking a pedagogically-disciplined approach to innovation with emergent technologies', *International Journal of Teaching and Learning in Higher Education*, 20, pp. 70–78.

Broersma, M. and Cutler, A. (2008) 'Phantom word activation in L2', *System*, 36(1), pp. 22–34.

Brown, J.D. and Kondo-Brown, K. (2006) 'Introducing connected speech' in Brown, J.D. and Kondo-Brown, K. (eds), *Perspectives on Teaching Connected Speech to Second Language Speakers*, pp. 1–15. Honolulu: University of Hawaii Press.

Brown, S. (2011) *Listening Myths*. Ann Arbor, MI: University of Michigan Press.

Buck, G. (2001) *Assessing Listening*. Cambridge: Cambridge University Press.

Burns, A. and Richards, J. (2009) *The Cambridge Guide to Second Language Teacher Education*. New York: Cambridge University Press.

Carrell, P., Dunkel, P. and Mollaun, P. (2002) *The Effects of Notetaking, Lecture Length and Topic on the Listening Component of TOEFL*. Princeton, NJ: Educational Testing Service.

Carroll, J.B. (1965) 'The prediction of success in foreign language training' in Glaser, R. (ed.), *Training, Research, and Education*, pp. 87–136. New York: Wiley.

Carroll, J.B. (1993) *Human Cognitive Abilities: A Survey of Factor-analytical Studies*. New York: Cambridge University Press.

Chafe, W. (1977) 'The recall and verbalization of past experience' in Cole, R. (ed.) *Current Issues in Linguistic Theory*. Norwood, NJ: Ablex.

Chalhoub-Deville, M. (1995) 'Deriving oral assessment scales across different tests and rater groups', *Language Testing*, 12, pp. 16–33.

Chang, A. (2007) 'The impact of vocabulary preparation on listening comprehension, confidence, and strategy use', *System*, 35, pp. 534–550.

Chang, A. (2009) 'Gains to L2 listeners from reading while listening vs listening only in comprehending short stories', *System*, 37, pp. 652–663.

Chang, A. and Read, J. (2007) 'Support for foreign language listeners: its effectiveness and limitation', *RELC Journal*, 38, pp. 375–394.

Cheng, W., Greaves, C. and Warren, M. (2005) 'The creation of a prosodically transcribed intercultural corpus: the Hong Kong Corpus of Spoken English (prosodic)', *ICAME Journal: Computers in English Linguistics*, 29, pp. 47–68.

Choe, Y. (2011) 'Effects of task complexity and English proficiency on EFL learners' task production in SMSC', *Multimedia-Assisted Language Learning*, 14, pp. 3–34.

Chung, J. (2002) 'The effects of using two advance organizers with video texts for the teaching of listening in English', *Foreign Language Annals*, 35(2), pp. 231–241.

Churchland, P. (2006) *Neurophilosophy at Work*. Cambridge: Cambridge University Press.

Clark, J.M. and Paivio, A. (1991) 'Dual coding theory and education', *Educational Psychology Review*, 3(3), pp. 149–170.

Clark, R., Nguyen, F. and Swellen, J. (2006) *Efficiency in Learning: Evidence-based Guidelines to Manage Cognitive Load.* San Francisco: Wiley.

Clement, J., Lennox, C., Frazier, L., Solórzano, H., Kisslinger, E., Beglar, D. and Murray, N. (2009) *Contemporary Topics*, 3rd edn. White Plains, NY: Pearson Education.

Cope, B. and Kalantzis, M. (2012) *Literacies.* Cambridge: Cambridge University Press.

Cornelius, T. and Owen-DeSchryver, J. (2008) 'Differential effects of full and partial notes on learning outcomes and attendance', *Teaching of Psychology*, 35, pp. 6–11.

Crabbe, M. (2007) 'Learning opportunities: adding value to tasks', *ELT Journal*, 61, pp. 118–125.

Crawford, M. and Ueyama, Y. (2011) 'Coverage and instruction of reduced forms in EFL textbooks', *The Language Teacher*, 35(4), pp. 55–61.

Creese, A. (2005) *Teacher Collaboration and Talk in Multilingual Classrooms.* Bristol: Multilingual Matters.

Crookall, D. (2010) 'Serious games, debriefing, and simulation/gaming as a discipline', *Simulation Gaming*, 41, pp. 898–920.

Cross, J. (2009) 'Effects of listening strategy instruction on news videotext comprehension', *Language Teaching Research*, 13, pp. 151–176.

Cross, J. (2011) 'Comprehending news videotexts: the influence of visual context', *Language Leanring and Technology*, 15, pp. 44–68.

Csikszentmihalyi, M. (2002) *Flow: The Classic Work on How to Achieve Happiness.* London: Rider.

Cutler, A. (2011) 'Competition dynamics of second-language listening', *Quarterly Journal of Experimental Psychology*, 64, pp. 74–95.

Cutler, A. (2012) *Native Listening.* Cambridge, MA: MIT Press.

Davies, P. and Rinvolucri, M. (1989) *Dictation: New Methods, New Possibilities.* Cambridge: Cambridge University Press.

De Jong, N. (2005) 'Can second language grammar be learned through listening? An experimental study', *Studies in Second Language Acquisition*, 27, pp. 205–234.

DeKeyser, R. (2009) 'Cognitive-psychological processes in second language learning' in Long, M. and Doughty, C. (eds) *The Handbook of Language Teaching.* New York: Wiley.

Derwing, T. (1996) 'Elaborative detail: help or hindrance to the NNS listener?' *Studies in Second Language Acquisition*, 18, pp. 283–297.

Dien, J., Michelson, C. and Franklin, M. (2010) 'Separating the visual sentence N400 effect from the P400 sequential expectancy effect: cognitive and neuroanatomical implications', *Brain Research*, 1355, pp. 126–140.

Donaldson, R. and Haggstrom, M. (2006) 'Wired for sound: Teaching listening via computers and the world wide web' in Donaldson, R. and Haggstrom, M. (eds) *Changing Language Education Through CALL*. London: Taylor and Francis.

Donato, R. (1994) 'Collective scaffolding in language learning' in *Vygotskian Approaches to Second Language Research*, pp. 33–56. Westport, CT: Ablex.

Dörnyei, Z. (2005) *The Psychology of the Language Learner: Individual Differences in Second Language Acquisition*. New York: Wiley.

Dörnyei, Z. and Hadfield, J. (2013) *Motivating Learning*. Harlow, UK: Pearson Education.

Ducasse, A. and Brown, A. (2009) 'Assessing paired orals: raters' orientation to interaction', *Language Testing*, 26, pp. 423–443.

Dudeney, G., Hockly, N. and Pegrum, M. (2013) *Digital Literacies*. Harlow, UK: Pearson Education.

Eckert, P. and McConnell-Ginet, S. (2003) *Language and Gender*. Cambridge: Cambridge University Press.

Elkhafaifi, H. (2005) 'The effect of prelistening activities on listening comprehension in Arabic learners', *Foreign Language Annals*, 38, pp. 505–513.

Ellis, D. (2008) 'Testing the limits of Levelt's loops with delayed auditory playback', Ph.D. Dissertation, Florida State University, http://diginole.lib.fsu.edu/.

Ellis, R. (1989) 'Classroom learning styles and their effect on second language acquisition: a study of two learners', *System*, 17, pp. 249–262.

Erkaya, N. (2005) 'Benefits of using short stories in the EFL context', *Asian EFL Journal*, 8, pp. 1–13.

Field, J. (1998) 'Skills and strategies: towards a new methodology for listening', *ELT Journal*, 52, pp. 110–118.

Field, J. (2008) 'Bricks or mortar: which parts of the input does a second language listener rely on?' *TESOL Quarterly*, 42, pp. 411–32.

Firth, A. and Wagner, J. (1997) 'On discourse, communication, and (some) fundamental concepts in SLA research', *Modern Language Journal*, 81, pp. 285–300.

Fisher, J. and Harris, (1973) 'Effects of note-taking and review on recall', *Journal of Educational Psychology*, 65, pp. 321–325.

Ford, C., Fox, B. and Thompson, S. (2002) 'Introduction' in Ford, C., Fox, B. and Thompson, S. (eds) *The Language of Turn and Sequence*, pp. 3–13. New York: Oxford University Press.

Frey, N. and Fisher, D. (eds) (2008) *Teaching Visual Literacy: Using Comic Books, Graphic Novels, Anime, Cartoons, and More to Develop Comprehension and Thinking Skills*. Thousand Oaks: Sage.

Frommer, J. (2006) 'Wired for sound: teaching listening via computers and the world wide web' in Donaldson, R. and Haggstrom, M. (eds) *Changing Language Education Through CALL*. New York: Taylor and Francis.

de la Fuente, M.J. (2002) 'Negotiation and oral acquisition of L2', *Studies in Second Language Acquisition*, 24, pp. 81–112.

Fujimoto, D. (2009) 'Listener responses in interaction: A case for abandoning the term, backchannel', *Osaka Jogakuin University Research Repository*, 37, pp. 35–54.

Gardner, H. (1983) *Frames of Mind: The Theory of Multiple Intelligences*. New York: Basic Books.

Gardner, H. (2011) *The Unschooled Mind.* New York: Basic Books.

Gardner, R. (1985) *Social Psychology and Second Language Learning: The Role of Attitudes and Motivation*. London: Edward Arnold.

Gardner, R., Tremblay, P. and Masgoret, A. (1997) 'Towards a full model of second language learning: an empirical investigation', *Modern Language Journal*, 81, pp. 344–362.

Gass, S. and Mackey, A. (2006) 'Input, interaction, and output: an overview', *AILA Review*, 19(1), pp. 3–17.

Gernsbacher, M. and Kaschak, M. (2003) 'Language comprehension' in Nadel, L. (ed.) *Encyclopedia of Cognitive Science*. New York: Wiley.

Ginther, A. (2002) 'Context and content visuals and performance on listening comprehension stimuli', *Language Testing*, 19, pp. 133–167.

Glickstein, L. (2007) *Speaker's Guide to Authentic Connection*. Woodacre, CA: Leeway Press.

Goh, C. (2000) 'A cognitive perspective on language learners' listening comprehension', *System*, 28, pp. 55–75.

Goldstein, B. (2009) *Working with Images*. Cambridge: Cambridge University Press.

Goodwin-Jones, R. (2012) 'Emerging technologies. Digital video revisited: storytelling, conferencing, remixing', *Language Learning and Technology*, 16, pp. 1–19.

Graesser, A. and Person, N. (1994) 'Question asking during tutoring', *American Educational Research Journal*, 31, pp. 104–137.

Graham, S. (2006) 'Listening comprehension: the learners' perspective', *System*, 34, pp. 165–182.

Grgurović, M. and Hegelheimer, V. (2007) 'Help options and multimedia listening: students' use of subtitles and the transcript', *Language Learning & Technology*, 11, pp. 45–66.

Griffiths, G. and Keohane, K. (2000) *Personalising Language Learning*. Cambridge: Cambridge University Press.

Gruba, P. (2006) 'Playing the videotext: a media literacy perspective on video-mediated L2 listening', *Language Learning & Technology*, 10, pp. 77–92.

Guilberg, M. (2010) 'Multilingual modality: communicative difficulties and their solutions in second-language use' in Streeck, J., Goodwin, C. and LeBaron, C. (eds) *Embodied Interaction: Language and Body in the Material World*. Cambridge: Cambridge University Press.

Guilloteaux, M. (2007) 'Motivating language learners: a classroom-oriented investigation of teachers' motivational practices and students' motivation', Ph.D. thesis, University of Nottingham.

Guilloteaux, M. and Dörnyei, Z. (2008) 'Motivating language learners: a classroom-oriented investigation of the effects of motivational strategies on student motivation', *TESOL Quarterly*, 42, pp. 55–77.

Hanford, M. (2010) *The Language of Business Meetings*. Cambridge: Cambridge University Press.

Hasan, A. (2000) 'Learners' perceptions of listening comprehension problems', *Language, Culture and Curriculum*, 13, pp. 137–153.

Haynes, J., Driver, J. and Rees, G. (2005) 'Visibility reflects dynamic changes of effective connectivity between V1 and fusiform cortex', *Neuron*, 46(5), pp. 811–821.

Hellerman, J. (2007) 'The development of practices for action in classroom dyadic interaction: focus on task opening', *Modern Language Journal*, 91, pp. 83–96.

Herron, C., Hanley, J. and Cole, S. (1995) 'A comparison study of two advance organizers for introducing beginning foreign language students to video', *Modern Language Journal*, 79, pp. 387–395.

Hicks, J., Marsh, R. and Cook, G. (2005) 'An observation on the role of context variability in free recall', *Journal of Experimental Psychology Learning Memory and Cognition*, 31, pp. 1160–1164.

Hjorland, B. (2010) 'The foundation of the concept of relevance', *Journal of the American Society for Information*, 6, pp. 217–237.

Holliday, A. (2007) *Doing and Writing Qualitative Research*. London: Sage. http://journals.sfu.ca/tesl/index.php/tesl/article/viewFile/677/508.

Hubbard, P. (2011) 'An invitation to CALL: foundations of computer-assisted language learning'. http://stanford.edu/~efs/callcourse/.

Hymes, D. (1964) 'Toward ethnographies of communicative events' in Giglioli, P. (ed.) *Language and Social Context*. Harmondsworth: Penguin.

Hymes, D. (2009) 'Ways of speaking' in Duranti, A. (ed.) *Linguistic Anthropology: A Reader*, pp. 158–171.

Iimura, H. (2007) 'The listening process: effects of types and repetition', *Language Education and Technology*, 44, pp. 75–85.

Imai, Y. (2010) 'Emotions in SLA: new insights from collaborative learning for an EFL classroom', *Modern Language Journal*, 94, pp. 278–292.

Indefrey, P. and Cutler, A. (2004) 'Prelexical and lexical processing in listening' in Gazzaniga, M. (ed.) *The Cognitive Neurosciences*, pp. 759–74, 3rd edition. Cambridge, MA: MIT Press.

Ishikawa, Y. (2012) 'The influence of learning beliefs in peer-advising sessions: Promoting independent language learning', *Studies in Self-Access Learning Journal*, 3(1), pp. 93–107.

Jacobs, G. and McCafferty, S. (2006) 'Connections between cooperative learning and second language learning and teaching' in McCafferty, S. Jacobs, G. and Iddings, A. (eds) *Cooperative Learning and Second Language Teaching*. Cambridge: Cambridge University Press.

Jensen, E. and Vinther, T. (2003) 'Exact repetition as input enhancement in second language acquisition', *Language Learning*, 53, pp. 373–428.

Johnson, M., Karen J., Mitchell, K. and Ankudowich, E. (2012) 'The cognitive neuroscience of true and false memories', *Nebraska Symposium on Motivation*, 58, pp. 15–52.

Jones, B., Liacer-Arrastia, S. and Newbill, P. (2009) 'Motivating foreign language students using self-determination theory', *Innovation in Language Learning and Teaching*, 3, pp. 171–189.

Jones, L. and Plass, J. (2002) 'Supporting listening comprehension and vocabulary acquisition with multimedia annotations', *Modern Language Journal*, 86, pp. 546–61.

Juffs, A. and Harrington, M. (2011) 'Aspects of working memory in L2 learning', *Language Teaching*, 44, pp. 137–166.

Kanaoka, Y. (2009) *Academic Listening Encounters: The Natural World*. Cambridge: Cambridge University Press.

Keddie, J. (2009) *Images – Resource Books for Teachers*. Oxford: Oxford University Press.

Kiewra, K., Benton, S., Risch, N. and Christensen, M. (1995) 'Effect of note-taking format and study technique on recall and relational performance', *Contemporary Educational Psychology*, 20, pp. 172–187.

Ko, Y. (2010) 'The effects of pedagogical agents on listening anxiety and listening comprehension in an English as a Foreign Language context'. All Graduate Theses and Dissertations. Paper 822.

Koester, A. (2010) *Workplace Discourse*. London: Continuum.

Kolb, D. (1985) *Learning Style Inventory*. Boston, MA: McBer and Company.

Kolb, A. and Kolb, D. (2006) 'Learning styles and learning spaces: A review of the multidisciplinary application of experiential learning theory in higher education' in Sims, R.R. and Sims, S. (eds) *Learning Styles and Learning*. New York, NY: Nova Science Publishers, pp. 45–91.

Krashen, S. and Terrell, T. (1983) *The Natural Approach: Language Acquisition in the Classroom*. Hayward, CA: Alemany Press.

Krashen, S., Terrell, T., Ehrman, M. and Herzog, M. (1984) 'A theoretical basis for teaching the receptive skills', *Foreign Language Annals*, 17, pp. 261–274.

Kraus, N. (1999) 'Speech sound perception, neurophysiology, and plasticity', *Internal Journal of Pediatric Otorhinolaryngology*, 47, pp. 123–129.

Kurzweil, J. (2003) 'Techniques of oral quizzing', *Internet TESL Journal*, 9, pp. 1–3.

Lamb, M. (2004) 'Integrative motivation in a globalizing world', *System*, 32, pp. 3–19.

Lazaraton, A. (2002) *A Qualitative Approach to the Validation of Oral Language Tests*. Cambridge: Cambridge University Press.

Levinson, S. (1996) 'Language and space'. *Annual Review of Anthropology*, 25, pp. 353–382.

Lightbown, P. and Spada, N. (1999) *How Languages are Learned*, 2nd edition. Oxford: Oxford University Press.

Lindsay, R. and Gorayska, B. (2004) 'Relevance, goal-management and cognitive technology' in Gorayska, B. and Mey, J. (eds) *Cognition and Technology: Co-existence, Convergence, and Co-evolution*, pp. 63–108. Amsterdam: John Benjamins.

Little, D. (2008) 'Language learner autonomy: some fundamental considerations revisited', *Innovations in Language Learning and Teaching*, 1, pp. 14–29.

LoCastro, V. (1987) 'Aizuchi: A Japanese conversational routine' in Smith, L. (ed.) *Discourse Across Cultures*. New York: Prentice Hall.

Long, M. (1996) 'The role of the linguistic environment in second language acquisition' in Ritchie, W. and Bhatia, T. (eds) *Handbook of Second Language Acquisition*. San Diego: Academic Press.

Long, P. and Ehrmann, S. (2005) 'Future of the learning space: breaking out of the box', *Educause Review*, 3, pp. 42–58.

Loughran, J. (2010) 'Reflection through collaboration action research and inquiry' in Jones, N. (ed.) *Handbook of Reflection and Reflective Inquiry*, pp. 399–413. New York: Springer.

Lovett, M., Reder, L. and Lebiere, C. (1999) 'Modeling working memory in a unified architecture: an ACT-R perspective' in Miyake, A. and Shah, P. (eds) *Models of Working Memory*, pp. 135–182. Cambridge, MA: Cambridge.

Luoma, S. (2004) *Assessing Speaking*. New York: Cambridge University Press.

Lustig, C., May, C. and Hasher, L. (2001) 'Working memory span and the role of proactive interference', *Journal of Experimental Psychology*, 130, pp. 199–207.

Lynch, T. (2006) 'Academic listening: marrying top and bottom' in Uso-Juan, E. and Martinez-Flor, A. (eds) *Current Trends in the Development and Teaching of the Four Language Skills*. Berlin: DeGruyter.

Lynch, T. (2009) *Teaching Second Language Listening*. Oxford: Oxford University Press.

Lynch, T. (2011) 'Academic listening in the 21st century: reviewing a decade of research', *Journal of English for Academic Purposes*, 10, pp. 79–88.

Lyster, R. and Mori, H. (2006) 'Interactional feedback and instructional counterbalance', *Studies in Second Language Acquisition*, 28, pp. 269–300.

Major, R., Fitzmaurice, S., Bunta, F. and Balasubramanian, C. (2002) 'The effects of non-native accents on listening comprehension: implications for ESL assessment', *TESOL Quarterly*, 36, pp. 173–190.

Maleki, A. (2010) 'Techniques to teach communication strategies', *Journal of Language Teaching and Research*, 1, pp. 640–646.

Maley, A. and Duff, A. (2005) *Drama Techniques in Language Learning*, 3rd edition. Cambridge: Cambridge University Press.

Marsden, E. (2006) 'Exploring input processing in the classroom: an experimental comparison of processing instruction and enriched input', *Language Learning*, 56, pp. 507–566.

Martin, R. (2007) *The Psychology of Humor: An Integrative Approach*. London: Elsevier.

McBride, K. (2011) 'The effect of rate of speech and distributed practice on the development of listening comprehension', *Computer Assisted Language Learning*, 24, pp. 131–154.

McCafferty, S., Jacobs, G. and Iddings, A. (2006) *Cooperative Learning and Second Language Teaching*. Cambridge: Cambridge University Press.

McCarthy, M. (2010) 'Spoken fluency revisited', *English Profile Journal*, 1, pp. 1–15.

Milne, P. (2007) 'Motivation, incentives and organisational culture', *Journal of Knowledge Management*, 11(6), pp. 28–38. Emerald Group Publishing Limited.

Miyake, A. and Shah, P. (1999) *Model of Working Memory: Mechanisms of Active Maintenance and Executive Control*. Cambridge: Cambridge University Press.

Monaghan, P. and Christiansen, M. (2010) 'Words in puddles of sound: modelling psycholinguistic effects in speech segmentation', *Journal of Child Language*, 37, pp. 545–564.

Moore, B. (2004) *An Introduction to the Psychology of Hearing*, 5th edition. Oxford: Elsevier.

Moore, P. (2011) 'Collaborative interaction in turn-taking: a comparative study of European bilingual (CLIL) and mainstream (MS) foreign language learners in early secondary education', *International Journal of Bilingual Education and Bilingualism*, 14, pp. 531–549.

Moore, R. (2007) 'Spoken language processing: piecing together the puzzle', *Speech Communication*, 49, pp. 418–435.

Morgan, B. and Clarke, M. (2011) 'Identity in second language teaching and learning' in Hinkel, E. (ed.) *Handbook of Research in Second Language Teaching and Learning*, Vol. 2, pp. 817–836. New York: Routledge.

Moser, J. and Rogers, G. (2005) 'The power of linking service to learning', *Tech Directions*, 2, p. 18.

Mumford, S. (2005) 'Using creative thinking to find new uses for realia', *Internet TESL Journal*, 11, pp. 1–3. http://iteslj.org/Techniques/Mumford-Relia.html.

Murphey, T. (1992) *Resource Books for Teachers: Music and Song*. Oxford: Oxford University Press.

Murphey, T. (2000) *Shadowing and Summarizing*, NFLRC Video #11. Honolulu: University of Hawaii, Second Language Teaching and Curriculum Center.

Murphy, T.M. and Cross, V. (2002) 'Should students get the instructor's lecture notes?', *Journal of Biological Education*, 36, pp. 72–75.

Musiek, F. and Baran, J. (2007) *The Auditory System: Anatomy, Physiology and Clinical Correlates*. Boston, MA: Allyn & Bacon.

Mynard, J. and Carson, L. (2012) *Advising in Language Learning: Dialogue, Tools and Context*. Harlow: Pearson Education.

Nakamura, M., Iwano, K. and Furui, S. (2008) 'Differences between acoustic characteristics of spontaneous and read speech and their effects on speech recognition performance', *Computer Speech and Language*, 22, pp. 171–184.

Nakatani, Y. (2010) 'Identifying strategies that facilitate EFL learners' oral communication: a classroom study using multiple data collection procedures', *Modern Language Journal*, 94, pp. 116–136.

Nation, P. (2008) *Teaching Vocabulary: Strategies and Techniques*. Boston: Heinle ELT.

Nation, P. and Newton, J. (2009) *Teaching ESL/EFL Listening and Speaking*. London: Routledge.

Neef, N., McCord, B. and Ferreri, S. (2006) 'Effects of guided notes versus completed notes during lectures on college students' quiz performance', *Journal of Applied Behavior Analysis*, 39, pp. 123–130.

Nishi, K. and Kewley-Port, D. (2007) 'Training Japanese listeners to perceive American English vowels: influence of training sets', *Journal of Speech, Language, and Hearing Research*, 50, pp. 1496–1509.

Norris, S. (2012) *Multimodality in Practice: Investigating Theory-in-Practice-Through-Methodology*. London: Routledge.

Norton, B. (2010) 'Language and identity' in Hornberger, N. and McKay, S. (eds) *Sociolinguistics and Language Education*. Bristol: Multilingual Matters.

Nunan, D. (1992) *Collaborative Language Learning and Teaching*. Cambridge: Cambridge University Press.

Ohala, J. (1996) 'Ethological theory and the expression of emotion in the voice', *Spoken Language ICSLP 96: Proceedings*. Fourth International Conference, Philadelphia, PA.

O'Malley, J. and Chamot, A. (1990) *Learning Strategies in Second Language Acquisition*. Cambridge: Cambridge University Press.

Oxford, R. (1990) *Language Learning Strategies: What Every Teacher Should Know*. Heinle ELT.

Oxford, R. (2011) *Teaching and Researching Language Learning Strategies*. Harlow: Pearson Education.

Paivio, A. (1986) *Mental Representations*. New York: Oxford University Press.

Paivio, A. (2007) *Mind and its Evolution: A Dual Coding Theoretical Interpretation*. Mahwah, NJ: Erlbaum.

Perfetti, C. and Lesgold, A. (1997) 'Discourse comprehension and sources of individual differences' in Just, M. and Carpenter, P. (eds) *Cognitive Processes in Comprehension*. Hillsdale, NJ: Erlbaum.

Pica, T. (2005) 'Classroom learning, teaching, and research: a task-based perspective', *Modern Language Journal*, 89, pp. 339–52.

Poldrack, R., Wagner, A., Prull, M., Desmond, J., Glover, G. and Gabrieli, J. (1999) 'Functional specialization for semantic and phonological processing in the left inferior prefrontal cortex', *NeuroImage*, 10, pp. 15–25.

Prashnig, B. (1998) *The Power of Diversity: New Ways of Learning and Teaching Through Learning Styles*. London: Continuum.

Prashnig, B. (2006) *Learning Styles in Action*. London: Network Continuum Publications.

Prince, P. (2012) 'Writing it down: issues relating to the use of restitution tasks in listening comprehension', *TESOL Journal*, pp. 65–86.

Puakpong, N. (2008) 'An evaluation of a listening comprehension program' in Zhang, F. and Barber, B. (eds) *Handbook of Research on Computer-Enhanced Language Acquisition and Learning*, pp. 275–293. Hershey, PA: IGI Global.

Purmensky, K. (2009) *Service-learning for Diverse Communities: Critical Pedagogy and Mentoring*. Charlotte, NC: Information Age Publishing.

Qiu, A. and Huang, J. (2012) 'The effects of dynamic image schema on ESL students' systematic improvement of listening comprehension: a dynamic system theory perspective', *International Journal of Learning and Development*, 2, pp. 241–252.

Rahimi, M. (2011) 'Use of syntactic elaboration techniques to enhance comprehensibility of EST texts', *English Language Teaching*, pp. 11–17.

Reidsma, D., Truong, K., van Welbergen, H., Neiberg, D. and Pammi, S., de Kok, I. and van Straalen, B. (2010) *Continuous Interaction with a Virtual Human*, pp. 24–38. Amsterdam: ENTERFACE 2010.

Rendaya, W. and Farrell, T. (2011) ' "Teacher, the tape is too fast!" Extensive listening in ELT', *ELT Journal*, 65, pp. 52–59.

Richards, K. (2006) ' "Being the teacher": Identity and classroom conversation', *Applied Linguistics*, 27, pp. 51–77.

Rickardts, J., Fajen, B., Sullivan, J. and Gillespie, G. (1997) 'Signalling, notetaking, and field independence-dependence in text comprehension and recall', *Journal of Educational Psychology*, 89, pp. 508–517.

Robinson, P. (2012) 'Individual differences, aptitude complexes, SLA processes, and aptitude test development' in Pawlak, M. (ed.) *New Perspectives on Individual Differences in Learning and Teaching*, Part II, pp. 57–75. Berlin: Springer.

Rogers, Y. and Price, S. (2009) 'How mobile technologies are changing the way children learn' in Durin, A. (ed.) *Mobile Technology for Children: Designing for Interaction and Learning*, pp. 3–22. Burlington, MA: Morgan Kaufman.

Rodgers, M. and Webb, S. (2011) 'Narrow viewing: the vocabulary in related television programs', *TESOL Quarterly*, 45, pp. 689–717.

Rogerson-Revell, P. (2010) 'Can you spell that for us nonnative speakers? Accommodation strategies in international business meetings', *Journal of Business Communication*, 47, pp. 432–454.

Romeo, K. and Hubbard, P. (2010) 'Pervasive CALL learner training for improving listening proficiency' in Levy, M., Blin, F. and Siskin, C. (eds) *WorldCall: International Perspectives on Computer-Assisted Language Learning*. New York: Routledge.

Rosch, E., Mervis, C., Gray, W., Johnson, D. and Boyes-Braem, P. (2004) 'Basic objects in natural categories' in Balota, D. and Marsh, E. (eds) *Cognitive Psychology: Key Readings*. New York: Psychology Press.

Rosova, V. (2007) 'The use of music in teaching English: diploma thesis'. Brno: Masaryk University, Faculty of Education, Department of English Language and Literature.

Ross, R. (1975) 'Ellipsis and the structure of expectation', *Occasional Papers in Linguistics*. Volume 1, November. San Jose State University.

Ross, S. and Berwick, R. (1992) 'The discourse of accommodation in oral proficiency interviews', *Studies in Second Language Acquisition*, 14, pp. 59–176.

Rost, M. (1990) *Listening in Language Learning*. London: Longman.

Rost, M. (2006) 'Areas of research that influence L2 listening instruction' in Usó-Juan, E. and Martínez-Flor, A. (eds) *Current Trends in the Development and Teaching of the Four Language Skills*. Amsterdam: Mouton de Gruyter.

Rost, M. (2011) *Teaching and Researching Listening*, 2nd edition. London: Pearson Education.

Rost, M. and Ross, S. (1991) 'Learner use of strategies in interaction: typology and teachability', *Language Learning*, 41, pp. 235–268.

Ruhe, V. (1996) 'Graphics and listening comprehension', *TESL Canada Journal*, 14, pp. 45–59.

Rumelhart, D. (1980) 'Schemata: the building blocks of cognition' in Spiro, R., Bruce, B. and Brewer, W.F. (eds) *Theoretical Issues in Reading Comprehension*. Hillsdale, NJ: Erlbaum.

Russell, N. (2007) 'Teaching more than English: connecting ESL students to their community through service learning', *Phi Delta Kappan*, 6, pp. 770–771. Bloomington, IN: Phi Delta Kappa International.

Ryokai, K., Vaucelle, C. and Cassell, J. (2003) 'Virtual peers as partners in storytelling and literacy learning', *Journal of Computer Assisted Learning*, 19, pp. 195–208.

Sakai, H. (2009) 'Effect of repetition of exposure and proficiency level in L2 listening tests', *TESOL Quarterly*, 43, pp. 360–372.

Salehzadeh, J. (2006) *Academic Listening Strategies: A Guide to Understanding Lectures*. Ann Arbor, MI: University of Michigan Press.

Sarangi, S. and Roberts, C. (2002) 'Discoursal (mis)alignments in professional gatekeeping encounters' in Kramsch, C. (ed.) *Language Acquisition and Language Socialisation: Ecological Perspectives*, London: Continuum, pp. 197–227.

Schmitt, N. (2000) *Vocabulary in Language Teaching*. Cambridge: Cambridge University Press.

Schmitt, N. (2008) 'Instructed second language vocabulary learning', *Language Teaching Research*, 12, pp. 329–363.

Schmitt, N. (2010) *Researching Vocabulary: A Vocabulary Research Manual*. Houndmills, UK: Basingstoke.

Schoepp, K. (2001) 'Reasons for using songs in the ESL/EFL classroom', *Internet TESL Journal*, 7, pp. 1–4.

Seedhouse, P. (2004) *The Interactional Architecture of the Language Classroom: A Conversation Analysis Perspective*. Oxford: Blackwell.

Segalowitz, N. (2008) 'Automaticity and second languages' in Long, M.H. and Doughty, C.J. (eds) *The Handbook of Language Teaching*. New York: Wiley.

Singleton, D. and Ryan, L. (2004) *Language Acquisition: The Age Factor*. Clevedon, UK: Multilingual Matters.

Snow, C. and Hoefnagel-Hohne, M. (1978) 'The critical period for language acquisition: evidence from second language learning', *Child Development*, 49, pp. 1114–1128.

Sperber, D. and Wilson, D. (2004) 'Relevance theory' in Ward, G. and Horn, L. (eds) *Handbook of Pragmatics*. Oxford: Blackwell, pp. 607–632.

Svalberg, A. (2009) 'Engagement with language: interrogating a construct', *Language Awareness*, 3, pp. 242–258.

Swain, M. (1993) 'The Output Hypothesis: just speaking and writing aren't enough', *Canadian Modern Language Review*, 50, pp. 158–64.

Swain, M. (2010) 'The inseparability of cognition and emotion in second language teaching', *Language Teaching*, 10, pp. 1–13.

Sydorenko, T. (2010) 'Modality of input and vocabulary acquisition', *Language Learning & Technology*, 14, pp. 50–69.

Tannen, D. (1993) 'What's in a frame? Surface evidence for underlying expectations' in Tannen, D. (ed.) *Framing in Discourse*. Oxford: Oxford University Press.

Tannen, D. (2001) *You Just Don't Understand: Women and Men in Conversation*. New York: HarperCollins.

Tardy, C. and Snyder, B. (2004) '"That's why I do it": flow and EFL teachers' practices', *ELT Journal*, 58, pp. 118–128.

Taylor, E. (2001) 'Transformative learning theory: a neurobiological perspective of the role of emotions and unconscious ways of knowing', *International Journal of Lifelong Education*, 20, pp. 218–236.

Thornbury, S. (2005) *How to Teach Speaking*. London: Pearson.

Thorne, S., Black, R. and Sykes, J. (2009) 'Second language use, socialization, and learning in internet interest communities and online gaming', *Modern Language Journal*, 93, pp. 802–821.

Thothathiri, M. and Snedeker, J. (2008) 'Give and take: syntactic priming during spoken language comprehension', *Cognition*, 108, pp. 51–68.

Tomlinson, B. (2001) 'The inner voice: a critical factor in L2 learning', *Journal of the Imagination in Language Learning and Teaching*, 6, pp. 1–10. http://www.njcu.edu/cill/vol6/tomlinson.html.

Tong, Y. and Tong, D. (2010) 'An empirical study on web-based English listening autonomous learning model', *Computer Science and Information Technology (ICCSIT)*, 2010 3rd IEEE International Conference, pp. 524–527.

Toohey, K. (2000) *Learning English in School: Identity, Social Relations and Classroom Practice*. Clevedon, UK: Multilingual Matters.

Ur, P. (1984) *Teaching Listening Comprehension*. Cambridge: Cambridge University Press.

Ushioda, E. (2010) 'Motivation and SLA: bridging the gap', *EUROSLA Yearbook*, 10, pp. 5–20.

Ushioda, E. (2008) 'Motivation and good language learners' in Griffiths, C. (ed.) *Lessons from Good Language Learners*, pp. 19–34. Cambridge: Cambridge University Press.

Ushioda, E. (2009) 'Motivating learners to speak as themselves' in Murray, G., Gao, X. and Lamb, T. (eds) *Identity, Motivation and Autonomy in Language Learning*. Bristol: Multicultural Matters.

Vandergrift, L. (1997) 'The comprehension strategies of second language (French) listeners: a descriptive study', *Foreign Language Annals*, 30, pp. 387–409.

Vandergrift, L. (2003) 'Orchestrating strategy use: toward a model of the skilled second language listening', *Language Learning*, 53, pp. 463–496.

Vandergrift, L. (2005) 'Relationships among motivation orientations, meta-cognitive awareness and proficiency in L2 listening', *Applied Linguistics*, 26, pp. 70–89.

Vandergrift, L. (2011) 'Second language listening: presage, process, product, and pedagogy' in Hinkel, E. (ed.) *Handbook of Research in Second Language Teaching and Learning*, 2, pp. 455–471. New York: Routledge.

Vandergrift, L. and Goh, C. (2012) *Teaching and Learning Second Language Listening: Metacognition in Action*. New York: Routledge.

Vandergrift, L. and Tafaghodtari, M. (2010) 'Teaching students how to listen does make a difference', *Language Learning*, 60, pp. 470–497.

Vanderplank, R. (2010) 'Déjà vu? A decade of research on language laborat-ories, television and video in language learning', *Language Teaching*, 43, pp. 1–37.

VanPatten, B., Inclezan, D., Salazar, H. and Farley, A. (2009) 'Processing instruction and dictogloss', *Foreign Language Annals*, 42, pp. 557–575.

Vasiljevic, Z. (2010) 'Dictogloss as an interactive method of teaching listen-ing comprehension', *English Language Teaching*, 3, pp. 41–52.

Via, R. (1976) *English in Three Acts*. Honolulu: University of Hawaii Press.

Villaume, W. and Bodie, G. (2007) 'Discovering the listener within us: the impact of trait-like personality variables and communicator styles on preferences for listening style', *International Journal of Listening*, 21, pp. 102–23.

Wagner, M. and Urios-Aparisi, E. (2008) 'Pragmatics of humor in the foreign language classroom: learning (with) humor' in Putz, M. and Neff-Vanverselaer, J. (eds) *Developing Contrastive Pragmatics: Interlanguage and Cross-cultural Perspectives*. Berlin: DeGruyter.

Wajnryb, R. (1990) *Grammar Dictation*. Oxford: Oxford University Press.

Wajnryb, R. (2003) 'Stories: narrative activities for the language classroom', *Cambridge Handbooks for Language Teachers*. Cambridge: Cambridge University Press.

Webb, S. (2010) 'A corpus driven study of the potential for vocabulary learning through watching movies', *International Journal of Corpus Linguistics*, 15(4), pp. 497–519.

Weinart, R. (1995) 'The role of formulaic language in second language acquisition: a review', *Applied Linguistics*, 16, pp. 180–205.

Wen, Z. and Skehan, P. (2011) 'A new perspective on foreign language aptitude research: building and supporting a case for working memory as language aptitude', *Ilha Do Desterro: A Journal of English Language, Literatures and Cultural Studies*, 60, pp. 15–44.

White, C. and Burgoon, J. (2006) 'Adaptation and communicative design: patterns of interaction in truthful and deceptive conversations', *Human Communication Research*, 27, pp. 9–37.

White, G. (2008) 'Listening and good language learners' in Griffiths, C. (ed.) *Lessons from Good Language Learners*. Cambridge: Cambridge University Press.

Wilberschied, L. and Berman, P. (2004) 'Effect of using photos from authentic video as advance organizers on listening comprehension of an EFL Chinese class', *Foreign Language Annals*, 37, pp. 534–540.

Williams, M. and Burden, R. (1999) 'Students developing conceptions of themselves as language learners', *Modern Language Journal*, 83, pp. 193–201.

Williams, M., Burden, R. and Lanvers, U. (2002) 'French is the language of love and stuff: student perceptions of issues related to motivation in learning a foreign language', *British Educational Research Journal*, 28, pp. 503–528.

Williams, M., Burden, R., Poulet, G. and Maun, I. (2004) 'Learners' perception of their successes and failures in foreign language learning', *Language Learning Journal*, 30, pp. 19–29.

Wilson, JJ (2008) *How to Teach Listening*. Harlow: Pearson Education.

Wilson, M. (2003) 'Discovery Listening – improving perceptual processes', *ELT Journal*, 35, pp. 325–334.

Winke, P., Gass, G. and Sydorenko, T. (2010) 'The effects of captioning videos used for foreign language listening activities', *Language Learning and Technology*, 14, pp. 65–86.

Wise, A., Speer, J., Marbouti, F. and Hsiao, Y. (2012) 'Broadening the notion of participation in online discussions: examining patterns in learners' online listening behaviors', *Instructional Science*, 5, pp. 1–21.

Wise, D. and Forrest, S. (2003) *Great Big Book of Children's Games*. New York: McGraw-Hill.

Wolf, J. (2008) 'The effects of backchannels on fluency in L2 oral task production', *System*, 36, pp. 279–294.

Woodall, B. (2010) 'Simultaneous listening and reading in ESL: helping second language learners read (and enjoy reading) more efficiently', *TESOL Journal*, 1, pp. 186–205.

Wright, A. (2003) *1000+ Pictures for Teachers to Copy*. London: Thomas Nelson Publishers.

Wright, T. and Bolitho, R. (2009) 'Trainer development', *ELT Journal*, 63, pp. 176–179.

Yang, R. (1993) 'A study of the communicative anxiety and self-esteem of Chinese students in relation to their oral and listening proficiency in English' (Doctoral dissertation) in *Dissertation Abstracts International*, 54, 2132A. Atlanta, GA: University of Georgia.

Yang, S. (2009) 'Using blogs to enhance critical reflection and community of practice', *Educational Technology and Society*, 12, pp. 11–21.

Yilmaz, Y. and Granena, G. (2010) 'The effects of task type in synchronous computer-mediated communication', *ReCALL*, 22, pp. 20–38.

Zhao, Y. (2005) 'The effects of listeners' control of speech rate on second language comprehension' in Zhao, Y. (ed.) *Research in Technology and Second Language Education: Developments and Directions*. Charlotte NC: Information Age Publication.

Zheng, D., Young, M.F., Wagner, M. and Brewer, B. (2009) 'Negotiation for action: English language learning in game-based virtual worlds', *Modern Language Journal*, 93, pp. 489–511.

Zwaan, R. (2004) 'The immersed experiencer: Embodied theory of language comprehension' in Ross, B. (ed.) *The Psychology of Learning and Motivation: Advances in Research and Theory*. Oxford: Elsevier.

Zwaan, R. (2006) 'The construction of situation models in narrative comprehension: an event-indexing model', *Psychological Science*, 6, pp. 292–297.